TIM

Time Series Analysis

E. J. HANNAN

Department of Statistics,
Australian National University,
Canberra

LONDON
CHAPMAN AND HALL

First published 1960
by Methuen & Co. Ltd
Reprinted 1962, 1967
First published as a Science Paperback 1967
by Chapman and Hall Ltd
11 New Fetter Lane, London EC4P 4EE
Reprinted 1975, 1977

Printed in Great Britain
at the University Press, Cambridge

ISBN 0 412 20480 0

Distributed in the U.S.A.
by Halsted Press, a Division
of John Wiley & Sons, Inc.
New York

Contents

General Editor's Preface

It is not so very long ago that up-to-date text-books on statistics were almost non-existent. In the last few decades this deficiency has largely been remedied, but in order to cope with a broad and rapidly expanding subject many of these books have been fairly big and expensive. The success of Methuen's existing series of monographs, in physics or in biology, for example, stresses the value of short inexpensive treatments to which a student can turn for an introduction to, or a revision of, specialized topics.

In this new Methuen series the still-growing importance of probability theory in its applied aspects has been recognized by coupling together Probability and Statistics; and included in the series are some of the newer applications of probability theory to stochastic models in various fields, storage and service problems, 'Monte Carlo' techniques, etc., as well as monographs on particular statistical topics.

M. S. BARTLETT

Author's Preface

The analysis of time series is usually studied after a course in the classical part of statistical theory has been completed. For that reason this book assumes a knowledge of mathematics and statistics approximately up to the level of Cramer's text *Mathematical Methods of Statistics*. In the author's experience, even then, the student has difficulty, especially with the spectral theory. This is due to the unavoidable introduction of infinite dimensional vector spaces. A heuristic introduction to the spectral theory has therefore been given early in the first chapter, which reduces the discussion to one concerning finite dimensional vector spaces. Proper proofs of the main theorems of the spectral theory are given later in the chapter and in the appendix, and throughout the book this is done for most of the results which are cited.

The remaining chapters deal with statistical inference. The most general model considered is a univariate time series consisting of a time dependent mean plus a stationary component. A discussion of multivariate series was not possible because of the limited size of the book. The restriction to stationarity is necessary since a theory for evolutive series hardly exists, and almost all methods in use at present depend upon the reduction of the series to something approaching stationary form by elementary statistical devices.

This book was written while the author was a fellow at the Australian National University and arose out of a series of lectures delivered at the University of Western Australia. The author has learned much about this subject from discussion with Dr G. S. Watson and Dr P. Whittle, and to them, and most especially to Professor P. A. P. Moran, who introduced him to the subject, he expresses his gratitude.

E. J. HANNAN

CHAPTER I

The Spectral Theory of Discrete Stochastic Processes

1. Introduction and probability background

The statistical problems, with which this book will be concerned, arise when the data come in the form of a time series; that is a succession of observations made at equidistant points in time or covering equal intervals of time. The corresponding probability model, usually called a discrete stochastic process, is a family of (real valued) random variables,[1] x_t, where t varies over the integers. The adjective discrete refers, of course, to the nature (topology) of the set over which t varies. By adding the word 'stochastic' we indicate that the process generating the observations is not purely deterministic but incorporates chance elements.

The nature of the process $\{x_t\}$ will often be specified by stating the joint distribution of every finite set of these random variables. For example one might say that each x_t is normally distributed with mean μ_t and variance σ_t^2, independently of all other members of the family, though this would be a rather uninteresting example from the point of view of the present theory. If the family $\{x_t\}$ contained only n members then we should represent it by means of an n dimensional space Ω upon which was defined a probability measure μ, equivalent to the joint distribution of the n variables. Each point ω in this space would correspond to a particular observation on the n variables, this observation being the set of co-ordinates $x_t(\omega)$ of

[1] We assume a familiarity with probability theory and statistics roughly equivalent to that contained in Cramer [1].

the point. Thus the typical variable x_t is now regarded as a function on the space Ω, the value of which at the point ω is obtained by projecting that point on to the tth co-ordinate axis. With the infinite family $\{x_t\}$ for t varying over the integers we are led to consider a space Ω each point in which now corresponds to a particular history or 'realization' (let us say) of the process, that is a particular infinite sequence of values for the random variables x_t. Thus Ω becomes a space each point ω in which is a function defined upon the integers. We again use $x_t(\omega)$ to indicate the tth co-ordinate variable, that is the function on Ω which allots to ω the value which that function ω assumes at the integer t.

There now arises the question, easily answered in the finite dimensional case, whether it is possible to determine a probability measure μ, on the space Ω, which will be defined on a sufficiently wide class of sets[1] $\mathscr{A}$ in such a way that we may state that the joint distribution of the random variables $x_{t_1}(\omega), x_{t_2}(\omega), \ldots, x_{t_n}(\omega)$ defined on Ω will be the same as that of the variables $x_{t_1}, x_{t_2}, \ldots, x_{t_n}$ given in the first place (for every n; $t_1, t_2, \ldots, t_n$). That this is so has been proved in Kolmogoroff [1] under the single, eminently reasonable, condition that the initial joint distributions of the x_t should be compatible. By this is meant that the marginal distribution of any subset of the set $x_{t_1}, \ldots, x_{t_n}$ should be the same as the prescribed distribution of this subset. Thus, to repeat, having introduced Ω, μ and $\mathscr{A}$, we may regard the variable x_t for fixed t, as a random variable, $x_t(\omega)$, defined on the space Ω, and the set $\{x_t(\omega)\}$, for fixed ω, as a particular realization of the stochastic process. Unless we have reason to emphasize the dependence of $x_t(\omega)$ upon ω we shall drop the argument variable ω. No ambiguity will result.

The class of stochastic processes we have defined is sufficiently general to encompass an extraordinarily wide range of phenomena. To make useful inferences about a piece of data to hand it is therefore necessary to split off some subclass of the class of all of these processes. A most useful subclass is that of processes called stationary. A process is called strictly stationary if the distribution

[1] The class of sets $\mathscr{A}$ will constitute a Borel field. See Halmos [1].

of the set

$$x_{t_1}, x_{t_2}, \ldots, x_{t_n}$$

is the same as that of the set

$$x_{t_1+h}, x_{t_2+h}, \ldots, x_{t_n+h}$$

for every n; $t_1, t_2, \ldots, t_n$ and h.

Thus strict stationarity requires, in particular, the absence of trends. This includes not merely trends in the mean values of the x_t (if these exist) but also in their variances, for example, or in the relation between successive terms of the series. One may think of a strictly stationary process as one whose probability laws do not change through time. Though it is doubtful if one ever meets such a process in practice (even according to the wider definition of stationarity we are soon to give) it is still true that a considerable range of phenomena can be brought, by reasonably elementary statistical devices, to a form approximating sufficiently closely to stationarity for valid inferences to be made. A similar situation is met already in classical parts of statistics, where the whole range of small sample tests based on the techniques of analysis of variance depend for their validity on the normality of the data or its reduction to approximately normal form by relatively simple transformations.

As we shall see below, almost all statistical methods for the analysis of time series at the present time are based upon the use of moments of the first and second order. The definition of stationarity we have given therefore is in one sense unnecessarily strict, since it certainly relates to much more general functions of the observations than first and second order moments. On the other hand it is, in another sense, too narrow as it does not assure us of the existence of the second order quantities of which we shall be constructing estimates. It is thus preferable to use the following definition of stationarity. We shall call a process stationary in the wide sense if

$$\text{(i)}\quad \mathscr{E}(x_t^2) = \int_\Omega x_t(\omega)^2\, d\mu < \infty \quad \text{for all } t$$

$$\text{(ii)}\quad \mathscr{E}(x_{s+t}x_s) = \int_\Omega x_{s+t}(\omega)x_s(\omega)\, d\mu = \gamma_t$$

depends only upon t and not upon s. We call γ_t the tth serial covariance.

We have here used the symbol $\mathscr{E}$ for 'expectation'. The condition (i) makes this definition narrower than that already given but in a way which, we may hope, hardly matters. On the other hand, once (i) is imposed (so that the expectation in (ii) has a meaning) (ii) is a less restrictive condition than that given initially. The initial requirement certainly implies (ii) while a sequence of independent random variables with constant variances, but with otherwise arbitrary distributions, is an example of a process stationary in the wide sense but not in the strict sense. We shall use the word stationary exclusively to mean 'wide sense stationary' in what follows. Processes which are not stationary will be called evolutive.

2. Spectral theory of circularly defined processes

The spectral theory of stationary discrete stochastic processes, which will be dealt with in section 3 of this chapter, is concerned with a certain method of analysis of such a process into simple constituents, the fundamental particles, as it were, out of which it is composed. This analysis is fundamental not only in the probability theory but also in the problems of inference treated below. In order to exhibit the essentials of this analysis freed from the technicalities forced upon us by the fact that we are concerned with an infinity of random variables, we shall consider in this section a finite dimensional case which though unreal leads to reality asymptotically. Some of the concepts introduced will prove useful later, in the sampling theory of serial correlations.

Essentially what we must do is to replace the integers by a finite set. It is necessary, however, to maintain the property of the integers of being an additive group. The simple case we are led to consider, therefore, is that of a process where t varies over the integers reduced modulo n. By this we mean the symbols; $0, 1, 2, \ldots, n-1$; with a new law of addition defined by the rule that we take the remainder after division by n. We shall call such a process 'circular'. Our stationary condition (ii) required that $\mathscr{E}(x_{s+t}x_s)$ should depend only upon t. When we translate our stationarity conditions into the present context we therefore get

(i) $\gamma_0 = \mathscr{E}(x_t^2) < \infty$

(ii) $\gamma_t = \mathscr{E}(x_{s+t}x_s) = \mathscr{E}(x_{s+t+1}x_{s+1}) = \ldots$

$$\ldots = \mathscr{E}(x_0 x_{n-t}) = \gamma_{n-t};\ t = 0, 1, \ldots, n-1.$$

The space Ω over which each random variable is defined is now of course n-dimensional.

We consider the operator U defined by

$$Ux_t(\omega) = x_{t+1}(\omega).$$

We may extend the definition of U to any linear combination of the variables $x_t(\omega)$ by

$$U\sum_0^{n-1} \alpha_t x_t(\omega) = \sum_0^{n-1} \alpha_t x_{t+1}(\omega)$$

where now it is convenient to allow the α_t to be complex numbers.

The set of all such linear combinations may be made into a vector space, $\mathscr{V}$, defining addition by

$$\sum \alpha_t x_t(\omega) + \sum \beta_t x_t(\omega) = \sum (\alpha_t + \beta_t) x_t(\omega)$$

and if we choose the $x_t(\omega)$ as the basic vectors of the space, $\mathscr{V}$, so that the space may be considered as a space of n-tuples,

$$\{\alpha_0, \alpha_1, \ldots, \alpha_{n-1}\},$$

of complex numbers, the matrical form of U^{-1} becomes

$$\begin{bmatrix} 0 & 1 & 0 & \ldots & 0 \\ 0 & 0 & 1 & \ldots & 0 \\ \cdot & \cdot & \cdot & & \cdot \\ \cdot & \cdot & \cdot & \ldots & \cdot \\ \cdot & \cdot & \cdot & & \cdot \\ 0 & 0 & 0 & & 1 \\ 1 & 0 & 0 & & 0 \end{bmatrix} \tag{1}$$

The operator U, and its infinite dimensional analogue, play a fundamental part in the theory for the (multiplicative) group formed from U and its powers (there are only n distinct powers in the circular case) describes precisely the symmetry introduced by the stationarity conditions.

We now introduce into $\mathscr{V}$ the inner product[1]

$$\left(\sum \alpha_t x_t, \sum \beta_t x\right) = \mathscr{E}\left\{\sum \alpha_t x^t \sum \bar{\beta}_t x\right\}.$$

This is evidently bilinear while if x, y are two elements of $\mathscr{V}$, $(x, y) = \overline{(y, x)}$, $(x, x) \geqslant 0$.

We shall use $\| x \|$ to indicate the norm (length) of the vector[2] $x \,\varepsilon\, \mathscr{V}$,

$$\| x \|^2 = (x, x).$$

With respect to this inner product the operator U becomes a unitary operator since

$$\left(U \sum \alpha_t x_t, U \sum \beta_t x_t\right) = \mathscr{E}\left\{\sum \alpha_t x_{t+1} \sum \bar{\beta}_t x_{t+1}\right\}$$
$$= \mathscr{E}\left\{\sum \alpha_t x_t \sum \bar{\beta}_t x_t\right\} = \left(\sum \alpha_t x_t, \sum \beta_t x_t\right).$$

The operator U is seen to have the proper values[3]

$$e^{-i\psi_j}; \quad \psi_j = 2\pi j/n; \quad j = 1, 2, \ldots, n$$

with corresponding proper vectors

$$\sum_{k=1}^{n} e^{ik\psi_j} x_k.$$

We shall retain this notation for the numbers $2\pi j/n$ throughout the book.

[1] The reader who is more familiar with a purely algebraic treatment of vector spaces and operators in terms of n-tuples and matrices may note that we may choose a basis in $\mathscr{V}$ so that the inner product of two vectors becomes the usual row-column product of the two sets of n-tuples which represent the two vectors in that basis.

[2] For ε read 'belongs to'.

[3] Equivalent terms are of course eigenvalue, characteristic value, etc. The notation and terminology used here follows closely that used in Halmos [2]. The fact that these are the proper values and vectors is easily checked from the matrical form (1).

Thus we may write

$$U = \sum_{j=1}^{n} e^{i\psi_j} E_j \tag{2}$$

where the E_j are perpendicular projections satisfying

$$E_j E_k = \delta_{j,k} E_j, \quad \sum_{j=1}^{n} E_j = I.$$

Here I is the identity operator; $Ix = x$ for all $x \varepsilon \mathscr{V}$.

The matrical form of the operator E_j in the basis $\{x_t\}$ is

$$[n^{-1} e^{i(p-q)\psi_j}]$$

where the element in the pth row and the qth column has been written in. E_j is evidently Hermitian which must be so for a perpendicular projection. E_j projects into the subspace of $\mathscr{V}$ spanned by the proper vector corresponding to $e^{i\psi_j}$.

We may now write

$$x_t = U^t x_0 = \sum_{j=1}^{n} e^{it\psi_j} E_j x_0$$

$$= \sum_{j=1}^{n} e^{it\psi_j} z_j \tag{3}$$

let us say, so that $z_j = n^{-1} \sum_{1}^{n} e^{-it\psi_j} x_t$.

We evidently have

$$(z_j, z_k) = (E_j x_0, E_k x_0) = 0, j \neq k$$

since E_j and E_k project on to mutually orthogonal subspaces.

Finally

$$\gamma_t = \mathscr{E}(x_t x_0) = \sum_{j=1}^{n} e^{it\psi_j}(z_j, x_0) = \sum_{j=1}^{n} e^{it\psi_j} f_j. \tag{4}$$

Since

$$(z_j, x_0) = (E_j x_0, x_0) = (E_j^2 x_0, x_0) = (E_j x_0, E_j x_0)$$

it follows that

$$f_j = (z_j, z_j) = \mathscr{E}(|z_j|^2).$$

Clearly

$$f_j = n^{-1} \sum_1^n \gamma_t e^{-it\psi_j}. \tag{5}$$

We shall now re-write these three fundamental relations in a way which will suggest their infinite dimensional generalizations. We introduce the functions:

$$E(\lambda) = \sum_{\psi_j \leqslant \lambda + \pi} E_j, \quad z(\lambda) = \sum_{\psi_j \leqslant \lambda + \pi} z_j, \quad F(\lambda) = \sum_{\psi_j \leqslant \lambda + \pi} f_j$$

where the summation in each case is over those j for which $\psi_j \leqslant \lambda + \pi$. The first of these is of course an operator-valued function of λ but since only finite sums are involved no difficulty arises in its definition.

The three relations given above may now be written, using Stieltjes integrals, as

The spectral decomposition of the operator

$$U = \int_{-\pi}^{\pi} e^{i\lambda}\, dE(\lambda). \tag{2)$'$(}$$

The spectral representation of the variable

$$x_t = \int_{-\pi}^{\pi} e^{it\lambda}\, dz(\lambda) \tag{3)$'$(}$$

with
$$\mathscr{E}\{|z(\lambda)|^2\} = F(\lambda)$$

$$\mathscr{E}\{(z(\lambda_1) - z(\lambda_2))\overline{(z(\lambda_3) - z(\lambda_4))}\} = 0, \quad \lambda_1 \geqslant \lambda_2 > \lambda_3 \geqslant \lambda_4.$$

The representation of the γ_t as the Fourier–Stieltjes coefficients of the spectral function

$$\gamma^t = \int_{-\pi}^{\pi} e^{it\lambda}\, dF(\lambda) \tag{4)$'$(}$$

If we use $dz(\lambda)$ as a shorthand symbol for $z(\lambda) - z(\lambda-)$ then we see from the relation, $z_j = n^{-1} \sum_{1}^{n} e^{-it\varphi_j} x_t$, that $dz(-\lambda) = \overline{dz(\lambda)}$, while $dz(0)$ and $dz(\pi)$ are real valued. Putting

$$z(\lambda) = \tfrac{1}{2}\{u(\lambda) - iv(\lambda)\}, \quad 0 < \lambda < \pi$$

we obtain

$$\begin{aligned} 0 &= \mathscr{E}\{dz(\lambda)\,dz(-\lambda)\} = \mathscr{E}\{dz(\lambda)^2\} \\ &= \tfrac{1}{4}\mathscr{E}\{du(\lambda)^2\} - \tfrac{1}{4}\mathscr{E}\{dv(\lambda)^2\} - \tfrac{1}{2}i\mathscr{E}\{du(\lambda)dv(\lambda)\}, \quad 0 < \lambda < \pi, \end{aligned}$$

the first equality deriving from the fact that $dE(\lambda)$ and $dE(-\lambda)$ project on to mutually orthogonal subspaces.

Thus, putting $dz(0) = du(0)$, $dz(\pi) = du(\pi)$ we obtain

$$\begin{aligned} &\mathscr{E}\{du(\lambda)^2\} = \mathscr{E}\{dv(\lambda)^2\} = 2dF(\lambda), \quad 0 < \lambda < \pi, \\ &\mathscr{E}\{du(0)^2\} = dF(0), \quad \mathscr{E}\{du(\pi)^2\} = dF(\pi) \\ &\mathscr{E}\{du(\lambda_1)dv(\lambda_2)\} = 0, \quad 0 \leqslant \lambda_1, \lambda_2 \leqslant \pi. \end{aligned}$$

Also, by writing $u(\lambda)$ and $v(\lambda)$ in terms of $z(\lambda)$ we see that

$$\mathscr{E}\{du(\lambda_1)du(\lambda_2)\} = \mathscr{E}\{dv(\lambda_1)dv(\lambda_2)\} = 0, \quad 0 \leqslant \lambda_1 \neq \lambda_2 \leqslant \pi.$$

Thus we may, as an alternative to (3)′ and (4)′, write

$$x_t = \int_0^\pi \cos t\lambda\, du(\lambda) + \int_0^\pi \sin t\lambda\, dv(\lambda) \tag{3''}$$

and

$$\gamma_t = \int_0^\pi \cos t\lambda\, dG(\lambda) \tag{4''}$$

where we have put

$$\begin{aligned} dG(\lambda) &= 2dF(\lambda), \quad 0 < \lambda < \pi \\ dG(0) &= dF(0), \quad dG(\pi) = dF(\pi). \end{aligned}$$

If we think of the process $\{x_t\}$ as a family of realizations we see from (3)″ that the typical member is a sum of sine curves ($\tfrac{1}{2}n$ in number if n is even, $\tfrac{1}{2}(n + 1)$ in number if n is odd) whose frequencies do not vary from realization to realization. The amplitude and phase of the sine curve with frequency λ is determined by the numbers $du(\lambda)$ and $dv(\lambda)$, these being random variables with mean squares $dG(\lambda)$. The increments $du(\lambda)$ are orthogonal to each other and to $dv(\lambda)$ (which are also orthogonal to each other). Thus $G(\lambda)$ can be regarded as a function which distributes the total mean square of $x_t(G(\pi) = \gamma_0)$ over the various frequencies. The individual sine curves are the 'fundamental particles' out of which the process is composed and they represent separate, 'independent', aspects of the data if we interpret 'independent' in terms of the orthogonality of random variables.

Through most of what follows we shall use the relations (3)′ and (4)′ rather than the equivalent (3)″ and (4)″, since this leads, usually, to a simpler analysis.

3. Spectral theory in the general case

As n is increased the circularity restriction which was imposed becomes less and less important and we approach the general situation. The vector space upon which U operates is now infinite dimensional of course and is spanned by the set $\{x_t(\omega)\}$. A suitable reformulation of (2)′ which we shall not discuss here, may be obtained[1] from which (3)′ and (4)′ derive. The relation (3)′ must now be interpreted carefully. For every λ, $z(\lambda)$ is a random variable, defined upon Ω. It has mean square $F(\lambda)$ and orthogonal increments (in the sense of the second relation below (3)′). The formula (3)′ is to be understood in the sense that we may find a decomposition of the interval $[-\pi, \pi]$ by means of points λ_j so that the approximating sum

$$\sum_1^m e^{it\lambda_j}\{z(\lambda_j) - z(\lambda_{j-1})\}$$

differs from the random variable x_t by a quantity whose mean square is arbitrarily small. That is

$$\mathscr{E}\left[\,\left|\,x_t - \sum_1^m e^{it\lambda_j}\{z(\lambda_j) - z(\lambda_{j-1})\}\,\right|^2\,\right]$$

may be made arbitrarily small by choosing a sufficiently fine sub-

[1] See Riesz and Nagy [1] p. 280. Of course (3)″ and (4)″ continue to hold also

division of the interval.[1] The interpretation of $F(\lambda)$ remains essentially the same. The process is thought of as being composed of oscillations, with now, however, the possibility that an oscillation of every frequency is represented. The function $F(\lambda)$ distributes the total mean square of the process over the various oscillations. If $F(\lambda)$ has a jump at a frequency λ_0 then this frequency will be represented in (3)′ with an amplitude and phase determined by a pair of orthogonal random variables having non-zero mean square $2dF(\lambda_0)$. In general, however, the part of γ_0 which is explained by frequencies in the infinitesimal range $d\lambda$ will itself be infinitesimal.

The third relation does not need reinterpretation, save for the fact that the non-decreasing function $F(\lambda)$ may now increase continuously as well as by jumps. This relation may also be derived directly from the properties of the γ_t. We shall give this derivation here both for completeness and because we shall need the theorem again in Chapter V. We have, for any n, ξ_j and $t_1, \ldots, t_n$

$$\sum\sum_{j,k=1}^{n} \xi_j \bar{\xi}_k \gamma_{t_j - t_k} = \mathscr{E}\left\{\left|\sum_{1}^{n} \xi_j x_{t_j}\right|^2\right\} \geqslant 0$$

so that the γ_t form what is called a positive sequence. It is evident that this is a necessary condition in order that a relation of the form (4)′ should hold, for some $F(\lambda)$ which is non-decreasing, for using (4)′ the last expression becomes

$$\int_{-\pi}^{\pi} \sum\sum_{1}^{n} \xi_j \bar{\xi}_k e^{i(t_j\lambda - t_k\lambda)}\, dF(\lambda) = \int_{-\pi}^{\pi} \left|\sum_{1}^{n} \xi_j e^{it_j\lambda}\right|^2 dF(\lambda) \geqslant 0.$$

It is also sufficient, however. To show this we form the function

$$F_n(\lambda) = \frac{1}{2\pi}\sum_{-n}^{n}{}' \gamma_t \frac{e^{it\pi} - e^{-it\lambda}}{it}\left(1 - \frac{|t|}{n}\right)$$

where the prime indicates that the term for $t = 0$ is $\gamma_0(\pi + \lambda)$. For

[1] A proof of the result (3)′, depending upon the proof of (4)′ which we shall give in this section, is given in the Appendix.

$\lambda_1 \geqslant \lambda_2$ we have

$$F_n(\lambda_1) - F_n(\lambda_2) = \frac{1}{2\pi}\sum_{-n}^{n} \gamma_t \frac{e^{-it\lambda_2} - e^{-it\lambda_1}}{it}\left(1 - \frac{|t|}{n}\right)$$

$$= \frac{1}{2\pi n}\sum\sum_{s,t=1}^{n} \gamma_{s-t}\frac{e^{-i(s-t)\lambda_2} - e^{-i(s-t)\lambda_1}}{i(s-t)}$$

$$= \frac{1}{2\pi n}\int_{\lambda_2}^{\lambda_1}\sum\sum \gamma_{s-t}e^{-i(s-t)\lambda}\, d\lambda \geqslant 0$$

because of the positiveness of the sequence γ_t. Finally

$$F_n(-\pi) = 0, \quad F_n(\pi) = \gamma_0.$$

Thus each $F_n(\lambda)$ is a non-decreasing function whose total variation is γ_0. We may extract a subsequence $F_{n_j}(\lambda)$ from this sequence which converges to a non-decreasing function $F(\lambda)$ (which again will have total variation γ_0) at all points of continuity of $F(\lambda)$[1] and which may be uniquely defined by requiring it to be continuous to the right.[2] Since

$$\int_{-\pi}^{\pi} e^{it\lambda} dF_n(\lambda) = \gamma_t\left(1 - \frac{|t|}{n}\right); \quad |t| \leqslant n$$

$$= 0; \qquad |t| > n$$

we obtain the limit (4)′ for each fixed t by letting n increase infinitely.

We may put[3]

$$F(\lambda) = F_1(\lambda) + F_2(\lambda) + F_3(\lambda) \tag{1}$$

where

(i) $F_1(\lambda)$ is absolutely continuous (a.c.), i.e.

$$F_1(\lambda) = \int_{-\pi}^{\lambda} f(\lambda)\, d\lambda.$$

(ii) $F_2(\lambda)$ is a jump function, being constant save for jumps at a finite or denumerable set of points.

[1] See Cramer [1] p. 60. [2] If there is a jump at $-\pi$ we transfer it to π.
[3] See for example Riesz and Nagy [1] p. 15 and p. 53. For an example of a function of the type of the (pathological) $F_3(\lambda)$ see Munroe [1] p. 193.

(iii) $F_3(\lambda)$ is continuous with a zero derivative almost everywhere (a.e.). This part does not appear to be meaningful observationally and we shall neglect it in what follows.

We call $F(\lambda)$ the spectral distribution function and if only $F_1(\lambda)$ is present in the decomposition (1) we say that the process has a continuous spectrum with spectral density $f(\lambda)$. If only $F_2(\lambda)$ is present we say that the process has a discrete spectrum. If both are present the spectrum is said to be mixed. We shall always use $f(\lambda)$ for the derivative of the absolutely continuous part of $F(\lambda)$.

We shall now give some examples of stationary processes, specifying them in terms of some $F(\lambda)$ of standard types. In the course of these examples we shall be led to form expressions of the type

$$y = \int_{-\pi}^{\pi} g(\lambda)\, dz(\lambda) \tag{2}$$

where $g(\lambda)$ is square integrable with respect to $dF(\lambda)$. These expressions are to be thought of as the limit of a sequence of approximating sums which converge to y in the sense that the mean square error involved in replacing y by the approximating sum will tend to zero.[1]

Examples

(1) The simplest specification which one can make for $F(\lambda)$ is to require that it be a.c. with $f(\lambda) = \gamma_0/(2\pi)$, whence

$$\gamma_t = \frac{\gamma_0}{2\pi}\int_{-\pi}^{\pi} e^{it\lambda}\, d\lambda \begin{cases} = 0 & t \neq 0 \\ = \gamma_0 & t = 0. \end{cases}$$

Evidently $z(\lambda)$ here satisfies $\mathscr{E}\{|\,z(\lambda)\,|^2\} = \dfrac{1}{2\pi}\gamma_0(\lambda + \pi)$. Thus x_t is a process of orthogonal random variables. A process of this form is said to have a uniform spectrum.

(2) $F(\lambda) = F_2(\lambda)$ with jumps at the p points λ_j, the jth jump being f_j. Since now $dz(\lambda)$ is null save at the points λ_j, we obtain, from (3)′ of

[1] See Riesz and Nagy [1] Chapter IX and Hille and Phillips [1] Chapter III. A discussion which covers the examples with which we shall be concerned is given in the Appendix.

the previous section,

$$x_t = \sum_j z_j e^{it\lambda_j}, \quad z_j = dz(\lambda_j)$$

with $$\mathscr{E}\{|z_j|^2\} = f_j; \quad \sum f_j = \gamma_0.$$

(3) Since any continuous function, with period 2π, may be approximated uniformly by a trigonometric polynomial it is natural to consider the, evidently very general, process for which $F(\lambda)$ is a.c. with

$$f(\lambda) = \frac{1}{2\pi}\sum_{-q}^{q} \gamma_j e^{ij\lambda} \geqslant 0.$$

The polynomial $\sum_{-q}^{q} \gamma_j z^{q+j}$ has $2q$ roots. If z_1 is such a root[1] then

$$\sum_{-q}^{q} \gamma_j \bar{z}_1^{-(q+j)} = \bar{z}_1^{-2q} \sum_{-q}^{q} \gamma_j \bar{z}_1^{q-j}$$

$$= \bar{z}_1^{-2q}\left\{\overline{\sum_{-q}^{q} \gamma_j z_1^{q-j}}\right\} = \bar{z}_1^{-2q}\left\{\sum_{-q}^{q} \gamma_j \bar{z}_1^{q+j}\right\} = 0.$$

Thus $\bar{z}_1^{-1}$ is also a root and we may group the $2q$ roots into pairs $(z_1, \bar{z}_1^{-1}), (z_2, \bar{z}_2^{-1}), \ldots, (z_q, \bar{z}_q^{-1})$. Thus

$$f(\lambda) = \frac{\gamma_q}{2\pi} e^{-iq\lambda} \prod_1^q (e^{i\lambda} - z_j)(e^{i\lambda} - \bar{z}_j^{-1})$$

$$= \frac{\gamma_q}{2\pi}(-)^q \prod_1^q \bar{z}_j^{-1} \prod_1^q | e^{i\lambda} - z_j |^2 = \frac{1}{2\pi}\alpha_q^2 \left| \prod_1^q (e^{i\lambda} - z_j) \right|^2$$

$$= \frac{1}{2\pi}\left| \sum_0^q \alpha_j e^{ij\lambda} \right|^2 .$$

[1] We assume of course that $\gamma_q \neq 0$ (since otherwise q would be reduced) so that $z = 0$ is not a root.

However, we might equally well permute any of the pairs $(e^{i\lambda} - z_j)$, $(e^{-i\lambda} - \bar{z}_j)$ and write $f(\lambda)$ in one of the 2^q alternative forms $(2\pi)^{-1}\left|\sum_0^q \alpha_j^{(r)} e^{ij\gamma}\right|^2$, $r = 1, \ldots, 2^q$ where the polynomial inside the modulus sign is formed from the product of the first members of every pair. Not all of these 2^q polynomials need be distinct.

For example if

$$f(\lambda) = \frac{1}{2\pi}\{2e^{i2\lambda} - 9e^{i\lambda} + 14 - 9e^{-i\lambda} + 2e^{i2\lambda}\}$$

then we may write

$$f(\lambda) = \frac{1}{2\pi}\left| e^{i2\lambda} - 3e^{i\lambda} + 2\right|^2 = \frac{1}{2\pi}\left| 2e^{i2\lambda} - 3e^{i\lambda} + 1\right|^2$$

corresponding to the choices $(e^{i\lambda} - 1)$, $(e^{i\lambda} - 2)$ and $(e^{i\lambda} - 1)$, $(2e^{i\lambda} - 1)$, the remaining two choices for this case giving the same polynomials as those shown.

Choosing a particular representation of the (at most) 2^q possibilities, and dropping the r suffix for convenience, we may define[1]

$$\xi(\theta) = \int_{-\pi}^{\theta} \left\{\sum_0^q \alpha_j e^{-ij\lambda}\right\}^{-1} dz(\lambda)$$

taking $g(\lambda)$ in (2) as $\sum_0^q \alpha_j e^{-ij\lambda}$, $\lambda \leqslant \theta$ and zero elsewhere in $[\pi, \pi]$.

Then

$$\mathscr{E}\{|\,d\xi(\theta)\,|^2\} = \frac{1}{2\pi}\,d\theta. \tag{3}$$

[1] We here assume that $\sum^q \alpha_j e^{-ij\lambda} \neq 0$. If it has a zero more care is required.

We may write[1]

$$x_t = \int_{-\pi}^{\pi} e^{it\lambda} \sum_{0}^{q} \alpha_j e^{-ij\lambda}\, d\xi(\lambda) = \sum_{0}^{q} \alpha_j \varepsilon_{t-j}$$

where $\varepsilon_t = \int_{-\pi}^{\pi} e^{it\lambda}\, d\xi(\lambda)$

which from example (1) and the relation (3) is a process of orthogonal random variables. A process x_t of this form is called a (finite) moving average.

Evidently to every one of the possible representations of $f(\lambda)$ as the square of the modulus of a polynomial, corresponds a different moving average. For the numerical example just given these would be

$$x_t = 2\varepsilon_t^{(1)} - 3\varepsilon_{t-1}^{(1)} + \varepsilon_{t-2}^{(1)}$$

and $$x_t = \varepsilon_t^{(2)} - 3\varepsilon_{t-1}^{(2)} + 2\varepsilon_{t-2}^{(2)}.$$

(4) $F(\lambda) = F_1(\lambda)$ with $f(\lambda) = K\left|\sum_{0}^{p} \beta_j e^{ij\lambda}\right|^{-2}$. Evidently no root of $\sum_{0}^{p} \beta_j z^j$ can lie on the unit circle since $f(\lambda)$ is integrable. If we consider $\sum_{0}^{p} \beta_j x_{t-j} = \varepsilon_t$ we have

$$\varepsilon_t = \int_{-\pi}^{\pi} \sum_{0}^{p} \beta_j e^{i(t-j)\lambda}\, dz(\lambda)$$

$$= \int_{-\pi}^{\pi} e^{it\lambda}\, d\xi(\lambda); \quad \xi(\theta) = \int_{-\pi}^{\theta} \sum_{0}^{p} \beta_j e^{-ij\lambda}\, dz(\lambda).$$

Since $\mathscr{E}\{|\, d\xi(\lambda)\,|^2\} = K d\lambda$ we see that ε_t is once more a process of

[1] See the Appendix.

orthogonal random variables. In this case x_t is called a (finite) autoregression. If all of the roots z_j of $\sum_0^p \beta_j z^j$ lie outside of the unit circle we may expand $\left\{\sum \beta_j z^j\right\}^{-1}$ in a series of positive powers of z which will converge uniformly in a region including the unit circle. Introducing this expression into

$$x_t = \int_{-\pi}^{\pi} e^{it\lambda}\left\{\sum_0^p \beta_j e^{-ij\lambda}\right\}^{-1} d\xi(\lambda)$$

we see that we may express x_t as an (infinite) moving average in the ε_{t-j} for $j \geqslant 0$.

However, as in example 3 the expression $\left|\sum_0^p \beta_j e^{ij\lambda}\right|^2$ may be written in up to 2^p different forms, $\left|\sum_0^p \beta_j^{(r)} e^{ij\lambda}\right|^2$; $r = 1, \ldots, 2^p$; corresponding to the 2^p different choices from the pairs of roots z_j, $\bar{z}_j^{-1}$. Of course only one of these results in a polynomial $\sum_0^p \beta_j^{(r)} z^j$, having all of its roots outside of the unit circle. One also will have all of its roots *inside* that circle and thus $\left\{\sum \beta_j^{(r)} z^j\right\}^{-1}$, for this r will have an expansion involving only negative powers of z. The corresponding representation of x_t will constitute a moving average of infinite extent in orthogonal random variables, $\varepsilon_t^{(r)}$, say, where now however only $\varepsilon_{t+j}^{(r)}$ for $j \geqslant 0$ will occur. In the general case where $\sum \beta_j^{(r)} z^j$ has roots both inside and outside of the unit circle the representation will cease to be 'one-sided' and will involve both future and past values of the $\varepsilon_t^{(r)}$.

Often, of course, prior information in the way of knowledge that

x_t is generated by a process such that it is a moving average of shocks, which occurred in the past and present but not in the future, will enable us to choose one out of this set of up to 2^p possible representations as relevant and thus one particular relation of the form $\sum_0^p \beta_j x_{t-j} = \varepsilon_t$, such that $\sum_0^p \beta_j z^j$ has no roots in or on the unit circle, as the form of autoregressive relation generating x_t.

We shall discuss this form of autoregressive model again later but it is evident that a model in which x_t is determined by its immediate past values (linearly) and by a new random shock independent of these past values has considerable appeal on prior grounds.

(5) $F(\lambda)$ absolutely continuous; $f(\lambda) = K \dfrac{\left|\sum_0^q \alpha_j e^{ij\lambda}\right|^2}{\left|\sum_0^q \beta_j e^{ij\lambda}\right|^2}$. Here one is led to a process satisfying a relation of the form

$$\sum_0^p \beta_j x_{t-j} = \sum_0^q \alpha_j \varepsilon_{t-j} \tag{4}$$

where the ε_t are orthogonal random variables. A similar identification problem arises here to that which arose in examples 3 and 4, there now being up to 2^{p+q} possible processes of the form (4), with the same $f(\lambda)$. Again prior information may allow us to identify one as the appropriate process.

(6) Let $x_t = e^{it\xi}$ where ξ is a random variable on $[-\pi, \pi]$ with *probability* distribution function $F(\lambda)$.[1] Then

$$\gamma_t = \mathscr{E}\{x_{s+t}\bar{x}_s\} = \mathscr{E}(e^{it\xi}) = \int_{-\pi}^{\pi} e^{it\lambda}\, dF(\lambda)$$

so that x_t has the assigned *spectral* distribution function $F(\lambda)$.

[1] We have so far considered only real valued processes. However, the reader should have no difficulty in modifying this example so that it is real valued and gives essentially the same result.

4. Processes in continuous time

As has already been said we shall be concerned throughout this book with the analysis of observations taken at (equidistant) discrete points of time. Nevertheless in many cases it may be more realistic to regard the underlying process generating the observations x_t as one proceeding continuously through time. In fact, in some cases, the data may have initially been recorded continuously and then, for the purposes of digital computation (as distinct from the use of some analogue device), sampled at successive equidistant points. It is therefore necessary to say something about processes in continuous time.

We are now concerned with a process $\{x_t(\omega)\}$ where t varies over the real numbers. Again ω lies in a space Ω on a suitable class of sets in which a probability measure μ is defined. From the point of view of further developments of the theory of processes in continuous time (with which, however, we shall not be concerned) it is necessary to impose some additional restriction on the process which amounts to the requirement that the probability measure μ can be determined in terms of the probabilities relating to the process x_t at a denumerable set of points t_j (say the set of all points on the real line having rational co-ordinates).[1] The restrictions (i) and (ii) of the first section are again imposed where now, since γ_t is a function of the continuously varying real variable t, we write $\gamma(t)$ for this serial covariance. Some restriction on $\gamma(t)$ must be imposed in order to prevent it from being too irregular a function of time. From a practical point of view one would be content with the requirement that $\gamma(t)$ should be a continuous function of time but measurabilty of the function (with respect to Lebesgue measure on the real line) will suffice if the additional restriction on μ, just mentioned, is imposed.

The theory then proceeds in a manner which is essentially the same as that in which the theory of sections 2 and 3 was developed. The analogue of the relation (3)′ of the second section is now

$$x_t = \int_{-\infty}^{\infty} e^{it\lambda}\, dz(\lambda)$$

where $z(\lambda)$ is a process of orthogonal increments with λ varying on

[1] See Doob [1] p. 50 et seq.

the whole real line, while (4)′ has its analogue in

$$\gamma_t = \int_{-\infty}^{\infty} e^{it\lambda}\, dF(\lambda)$$

where $$\mathscr{E}\{|\, dz(\lambda)\,|^2\} = dF(\lambda).$$

The reason for the difference in the range of variation of λ in the two cases (discrete and continuous time) is evident. In the case of a process observed at discrete intervals any oscillation with frequency outside of the range from $-\pi$ to π is indistinguishable from some oscillation with frequency inside this range. Thus

$$\cos(\theta + 2k\pi)t = \cos\theta t$$

(t an integer) for $k = 0, \pm 1, \pm 2, \ldots$ When t varies continuously this is no longer so of course.

5. Prediction theory

We shall here give only a short account of prediction theory, proofs of the assertions made being omitted. This theory is concerned with the problem of predicting x_t, when the whole past of the realization is available, by means of a linear operation on the past values. We do not include this account because the prediction problem is directly relevant to our subsequent treatment of time series, for in all cases we shall be concerned with only a finite part of a realization, but rather because of the insight which it gives into the structure of a process.

The mean square error of prediction of x_t by means of a (finite) linear combination of past values; $x_{t_1}, x_{t_2}, \ldots, x_{t_n}$; $t_j < t$; is of the form

$$\mathscr{E}\left\{\left| x_t - \sum_{j=1}^{n} \alpha_j x_{t_j} \right|^2\right\} = \mathscr{E}\left\{\left| \int_{-\pi}^{\pi} e^{it\lambda} - \sum_{1}^{n} \alpha_j e^{it_j\lambda}\, dz(\lambda) \right|^2\right\}$$

$$= \int_{-\pi}^{\pi} \left| 1 - \sum_{1}^{n} \alpha_j e^{i(t_j - t)\lambda} \right|^2 dF(\lambda). \qquad (1)$$

The (minimized) error of prediction which we seek is then the infimum of this expression taken over the family of all such linear combinations.

It has been shown by Kolmogoroff [2] that

$$\sigma^2 = \inf \int_{-\pi}^{\pi} \left| 1 - \sum_{1}^{n} \alpha_j e^{-it_j\lambda} \right|^2 dF(\lambda)$$

$$= \exp \left\{ \frac{1}{2\pi} \int_{-\pi}^{\pi} \log 2\pi f(\lambda) \, d\lambda \right\} \tag{2}$$

a most remarkable formula. The right-hand side of (2) is to be interpreted as zero if $\int_{-\pi}^{\pi} \log f(\lambda) \, d\lambda = -\infty$ (the only possibility of divergence since $\log f(\lambda) < f(\lambda)$).[1] If $\sigma^2 = 0$ we say that the x_t process is deterministic. This is a very special use of the term since the process of example 6 of section 3 is deterministic in the 'every day' use of this term but need not be deterministic in our *linear* prediction sense, as (2) above shows. If $\sigma^2 > 0$ we say that the process is non-deterministic.

In case $\sigma^2 > 0$ we may decompose the process x_t as follows (Wold [1])

$$x_t = u_t + v_t = \sum_{0}^{\infty} \alpha_j \varepsilon_{t-j} + v_t \tag{3}$$

where, (1) the ε_t are mutually orthogonal and have mean square σ^2,

(2) $\alpha_0 = 1, \sum_{0}^{\infty} \alpha_j^2 < \infty$,

(3) $\mathscr{E}(\varepsilon_s v_t) = 0$ all s, t,

(4) the v_t process is deterministic.

Moreover the u_t process has an absolutely continuous spectral

[1] If $f(\lambda) = 0$ on a set whose measure is not zero, when the integral has no meaning, we still put the right-hand side equal to zero.

function with spectral density $f(\lambda)$, the density of the x_t process, with

$$f(\lambda) = \frac{\sigma^2}{2\pi}\left|\sum_0^\infty \alpha_j e^{ij\lambda}\right|^2$$

while the v_t process has spectral distribution function

$$F_v(\lambda) = F_2(\lambda) + F_3(\lambda)$$

where these last two are the jump and singular components of $F(\lambda)$.

If $\sigma^2 > 0$ and the v_t component is absent, so that $F(\lambda)$ is absolutely continuous and $\log f(\lambda)$ is integrable, then $x_t = \sum_0^\infty \alpha_j \varepsilon_{t-j}$, is said to be purely non-deterministic. Such a situation is clearly of a central nature for one feels intuitively that σ^2 will not be zero for most processes met in practice, while one hopes that one will be able to remove the deterministic component by regression methods.[1]

Of course $f(\lambda)$ may, in any case, be written in the form

$$f(\lambda) = \{f(\lambda)^{\frac{1}{2}}\}^2 = \left|\sum_{-\infty}^{\infty} \alpha_j e^{ij\lambda}\right|^2$$

where $\sum_{-\infty}^{\infty} \alpha_j e^{ij\lambda}$ is the Fourier series converging (in mean square) to the square integrable function $\{f(\lambda)\}^{\frac{1}{2}}$ (the positive square root of $f(\lambda)$ being taken). Thus, if $F(\lambda) = F_1(\lambda)$ we may in any case write

$$x_t = \sum_{-\infty}^{\infty} \alpha_j \varepsilon_{t-j}$$

where the ε_t form an orthogonal sequence. However, if $\sigma^2 > 0$ (and $F(\lambda) = F_1(\lambda)$) the results just stated show that this moving average is one-sided.

It may help the reader in understanding the meaning of formula (2) if the circular process of section 2 is considered. Here prediction is meaningless but the problem of interpolating, that is of predicting

[1] See Chapter V.

x_t given *all* of the remainder of the realization, may be considered. The error of interpolation then is

$$\sigma_i^2 = \inf \mathscr{E}\left\{\left| x_t - \sum_1^{n-1} \alpha_j x_{t-j} \right|^2\right\} \tag{4}$$

$$= \inf. \sum_{k=0}^{n-1} \left| 1 - \sum_{j=1}^{n-1} \alpha_j e^{-ij\psi_k} \right|^2 f_k.$$

If the f_k are all non-zero this sum cannot be zero since the vectors of n-tuples

$$(1, e^{i\psi_k}, e^{i2\psi_k}, \ldots, e^{i(n-1)\psi_k}), \quad k = 0, 1, \ldots, n-1$$

are linearly independent. If, however, one or more f_k is zero the corresponding component may be removed from all of the vectors so that they become n vectors in a space of dimension less than n and therefore linearly dependent. Thus the condition $f_k \neq 0$, which corresponds here to the condition $\log f(\lambda)$ integrable, is seen to be necessary and sufficient for the interpolation error to be different from zero.

If $\sigma_i^2 > 0$ its value is easily determined for, using $\hat{\alpha}_j$ for the coefficients for which the infimum in (4) is attained, we must have

$$\sum_{k=0}^{n-1} f_k e^{-ij\psi_k}\left(1 - \sum_{t=1}^{n-1} \hat{\alpha}_t e^{it\psi_k}\right) = 0, \quad j = 1, \ldots, n-1. \tag{5}$$

This follows from the fact that

$$x_t - \sum_{j=1}^{n-1} \hat{\alpha}_j x_{t-j}$$

must be orthogonal to x_{t-j} $(j = 1, \ldots, n-1)$ for the infimum to be reached.

The solution of (5) is

$$1 - \sum_{t=1}^{n-1} \hat{\alpha}_t e^{it\psi_k} = \frac{K}{f_k}$$

from which we obtain, by summing over k,

$$K = n\left(\sum_k \frac{1}{f_k}\right)^{-1}$$

Thus

$$\sigma_i^2 = \sum_k \left| 1 - \sum_{t=1}^{n-1} \hat{\alpha}_t e^{it\psi_k} \right|^2 f_k$$

$$= \sum_k \left(\frac{K}{f_k}\right)^2 f_k = n^2 \left(\sum_k \frac{1}{f_k}\right)^{-1}.$$

Putting $f(\lambda_j) = nf_j/(2\pi)$, where λ_j is $- \quad + \psi_j$, we derive, heuristically, the result

$$\sigma_i^2 = 1 \Big/ \frac{1}{2\pi}\int_{-\pi}^{\pi} \frac{d\lambda}{2\pi f(\lambda)}$$

which is in fact the error of interpolation in the general (non-circular) situation. If x_t is generated by a process with absolutely continuous spectral function and we know its mean value then our estimate of x_t, knowing neither past nor future, will have to be this mean value and the variance of estimate will be γ_0 (where γ_0 corresponds to the mean-corrected process). Thus we have the interesting array of results in this case:

Estimation situation	*Variance of error of prediction of x_t*
(1) No knowledge	Arithmetic mean of $2\pi f(\lambda)$
(2) Knowledge of all past	Geometric mean of $2\pi f(\lambda)$
(3) Knowledge of past and future	Harmonic mean of $2\pi f(\lambda)$

All of these means are of course with respect to the probability measure induced by $\frac{1}{2\pi}d\lambda$ on $[-\pi, \pi]$.

Kolmogoroff's formula for the error of prediction may be used to establish a connection between the spectral density of a process having an absolutely continuous spectral function and the proper values of the covariance matrix of $x_1, \ldots, x_n$[1] (see Whittle [1]). We shall find this result useful in Chapter V. We follow Grenander and Rosenblatt [1]. We consider the case where

$$|f(\lambda)| \leqslant \frac{M}{2\pi} < \infty$$

and denote by $\lambda_{j,n}$ the proper values of the covariance matrix

$$\Gamma_n = \begin{bmatrix} \gamma_0 & \gamma_1 & \cdots & \gamma_{n-1} \\ \gamma_1 & \gamma_0 & \cdots & \gamma_{n-2} \\ \cdot & \cdot & \cdot & \cdot \\ \cdot & \cdot & \cdot & \cdot \\ \cdot & \cdot & \cdot & \cdot \\ \gamma_{n-1} & \gamma_{n-2} & \cdots & \gamma_0 \end{bmatrix}. \tag{6}$$

It is well known that the $\lambda_{j,n}$ are real and positive and the inequality

$$\sum_{i,\,j=1}^{n}\sum \gamma_{i-j}\xi_i\bar{\xi}_j = \int_{-\pi}^{\pi} \left| \sum_{1}^{n} \xi_t e^{it\lambda} \right|^2 f(\lambda)\, d\lambda \leqslant M \sum_{1}^{n} |\xi_t|^2$$

shows that $|\lambda_{j,n}| \leqslant M$.

For $|\alpha| < M^{-1}$ we see that

$$0 < g(\lambda) = \frac{1}{2\pi}\{1 + 2\pi\alpha f(\lambda)\} < \frac{1}{\pi}$$

so that $g(\lambda)$ may be considered as the spectral density of a *purely non-deterministic* process, y_t.

It is easy to show that the minimum of

$$\mathscr{E}\left\{\left| y_t - \sum_{1}^{n} \beta_j y_{t-j} \right|^2\right\}$$

with respect to $\beta_1, \ldots, \beta_n$, is given by

$$\sigma_n^2 = |\mathbf{T}_{n+1}| / |\mathbf{T}_n|$$

where[2] $\mathbf{T}_n = \mathbf{I}_n + \alpha\Gamma_n$ (see Cramer [1] p. 305).

[1] The formula (I.2.5) shows that, for the circular process, the proper values of Γ_n are $2\pi\left(\frac{n}{2\pi}f_j\right) \sim 2\pi f(-\pi + \psi_j)$. The following discussion establishes this result more precisely.

[2] We use $\mathbf{I}_n$ to indicate the $(n \times n)$ identity matrix. $|\mathbf{T}_n|$ is the determinantal value of $\mathbf{T}_n$.

Thus
$$| \mathbf{T}_n | = \prod_1^n (1 + \alpha\lambda_{j,n}).$$

However, we know that σ_n^2 is a non-increasing function of n bounded below by

$$\lim_n \sigma_n^2 = 2\pi \exp\left\{\frac{1}{2\pi}\int_{-\pi}^{\pi} \log g(\lambda)\, d\lambda\right\}.$$

Thus

$$\begin{aligned}\frac{1}{2\pi}\int_{-\pi}^{\pi} \log\{1 + 2\pi\alpha f(\lambda)\}\, d\lambda &= \lim_{n\to\infty} \log \sigma_n^2 = \lim_{n\to\infty} \frac{1}{n}\sum_1^n \log \sigma_j^2 \\ &= \lim_{n\to\infty} \frac{\log|\mathbf{T}_{n+1}| - \log|\mathbf{T}_1|}{n} \\ &= \lim_{n\to\infty} \frac{1}{n}\sum_1^n \log(1 + \alpha\lambda_{j,n}).\end{aligned}$$

Using $\log(1 + x) = \sum_1^\infty (-)^{p-1}\frac{x^p}{p}$, which converges uniformly for $|x| \leqslant 1 - \delta$, $\delta > 0$, we have,

$$\lim_{n\to\infty} \sum_{p=1}^\infty (-)^{p-1}\frac{\alpha^p}{p} m_{n,p} = \sum_{p=1}^\infty (-)^{p-1}\frac{\alpha^p}{p}\frac{1}{2\pi}\int_{-\pi}^{\pi} \{2\pi f(\lambda)\}^p\, d\lambda$$

where we have put $m_{n,p} = \frac{1}{n}\sum_1^n \lambda_{j,n}^p$. It is easily seen that the left-hand side is bounded uniformly if

$$|\alpha| < (2M)^{-1} \left(\text{by } \sum \frac{|\alpha M|^p}{p} < \sum \frac{1}{p}\frac{1}{2^p} < \infty\right)$$

and it follows from Vitali's Theorem (Titchmarsh [1] p. 168) that the limit is uniform in α for α in this interval. Thus

$$\lim_{n\to\infty} m_{n,p} = \frac{1}{2\pi}\int_{-\pi}^{\pi} \{2\pi f(\lambda)\}^p\, d\lambda \quad p = 0, 1, \ldots$$

Thus if we consider the $\lambda_{j,n}$ as defining a sequence of distributions of functions $\lambda_n(x)$ of a random variable x distributed uniformly on the interval from $-\pi$ to π, ($\lambda_n(x)$ taking the value $\lambda_{j,n}$ in the interval $[-\pi + \psi_{j-1}, -\pi + \psi_j]$) we see that the distribution of $\lambda_n(x)$ converges to that of the function $2\pi f(x)$. Thus an approximation to the proper values $\lambda_{j,n}$ could, for large n, be obtained from the ordinates of $2\pi f(\lambda)$ at the points $-\pi + \psi_j$; $j = 1, \ldots, n$.

CHAPTER II

Estimation of the Correlogram and of the Parameters of Finite Parameter Schemes

1. Problems of inference in time series analysis

The classical part of the theory of estimation is concerned with the situation where a fixed, finite, number of independent observations is made. For this situation a large body of readily applicable general theory is available in the form of the central limit theorem and of the known optimal properties of the method of maximum likelihood. The former serves at least to give an asymptotic approximation to the distribution of a chosen estimator in a wide range of situations, while the latter provides a solution to the problem of choosing an estimate with certain optimal properties (including asymptotic normality).

Considerable extensions have been made in the central limit theorem which widen the domain of its applicability to that of dependent random variables. Some additional restrictions (over and above those which are analogues of requirements in the classic case) on the nature of the dependence has of course to be imposed and this (generally) takes the form of a requirement that the dependence between terms in the sequence should eventually fall off sufficiently quickly with their distance apart. In relation to the theory presented in Chapter I this has the unfortunate feature of relating the additional restriction to independence whereas the results of the first chapter relate to orthogonality. That it will be impossible to attain a form of the central limit theorem by imposing restrictions merely upon the nature of the cross product between terms far enough apart is evident from the fact that one can define a stationary sequence of *orthogonal* random variables which does not obey the central limit theorem (see Grenander and Rosenblatt [1] p. 180). One should not, however, make too much of this difficulty for the spectral theory assumes such importance in the theory of time series analysis because it is possible to do so much merely with second order quantities (and because the use of moments of higher order introduces such intractable difficulties) so that the necessary theory is 'tailored' to fit the methods used. It is not to be thought that orthogonality is a more basic concept than independence. Indeed one feels intuitively, let us say, that the motion of the waves which beat upon the seashores of today is totally unrelated to and not merely uncorrelated with that of those which beat some age ago. We shall here quote a form of central limit theorem for dependent random variables due to Diananda [1]. We call a sequence of random variables, y_t, m-dependent if, for some integer m, y_s and y_t are independent whenever $|s - t| > m$. We consider a stationary process of the form

$$x_t = \sum_0^\infty \alpha_j y_{t-j}$$

where (1) $\sum_0^\infty |\alpha_j| < \infty$.

(2) The process $\{y_t\}$ is strictly stationary, is m-dependent and has zero mean and finite variance. Then x_t is also strictly stationary with zero mean and finite variance while the distribution of

$$\frac{1}{\sqrt{n}} \sum_1^n x_t$$

tends to the normal distribution with zero mean and variance $\sum_{-\infty}^{\infty} \gamma_t$ where $\gamma_t = \mathscr{E}\{x_{s+t}x_s\}$.

In a situation where the process generating the observations is prescribed (including the distribution of the x_t) save for a finite set of parameters, $\theta_1, \theta_2, \ldots, \theta_p$, the likelihood equations could be written down and solved for the estimators $\hat{\theta}_j$ and Wald [1] has shown that, under reasonable regularity conditions, these $\hat{\theta}_j$ will be consistent and asymptotically unbiased with (asymptotically optimal) covariance properties which are the direct analogues of those for the case of independent observations. This is, however, a rather unlikely circumstance. An alternative procedure is to use the likelihood equations, written down on the basis of a *normal distribution* (or some modification of these equations which neglects terms which are asymptotically negligible), to obtain those estimates which would usually be called 'least squares' estimates. The asymptotic distribution properties of these estimates, free from any assumption about normality of distribution, can then be derived under fairly general circumstances. We shall discuss this in later sections of this chapter. However, we shall first consider the estimation of the mean of x_t and the constants γ_t.

2. Ergodic theory

We shall confine ourselves here to the 'statistical ergodic theorem' for stationary processes and shall merely mention that if the process is strictly stationary stronger results are available.

Let us consider a stationary process $\{x_t\}$ for which $\mathscr{E}(x_t) \equiv 0$. If the mean is unknown it is natural to estimate it, given n consecutive observations on the process, by

$$\bar{x} = \frac{1}{n}\sum_{1}^{n} x_t.$$

However, the justification for this procedure is not as obvious as might appear for we have, effectively, *one* observation on a vector random variable

$$\{x_1(\omega), x_2(\omega), \ldots, x_n(\omega)\}$$

and are expecting this one observation to tell us very accurately, if the dimension of the space is large enough, what the common mean of the elements of the vector is. The justification for the procedure constitutes the statistical ergodic theorem.

We may write

$$\bar{x} = \int_{-\pi}^{\pi} \frac{1}{n} \sum_{1}^{n} e^{it\lambda} \, dz(\lambda).$$

Since

$$\left| \frac{1}{n} \sum_{1}^{n} e^{it\lambda} \right|^2 = \frac{(\sin \frac{1}{2}n\lambda)^2}{(n \sin \frac{1}{2}\lambda)^2}$$

converges boundedly to the function which is zero save at $\lambda = 0$, where it is unity, this expression converges in the mean to $z(0) - z(0-)$. Thus $\bar{x}$ converges in the mean[1] to a random variable, which will, however, not be zero with probability one, unless $dF(0)$ is zero. The importance of this failure of $\bar{x}$ to converge to zero depends, of course, upon the nature of the situation. Situations can, of course, arise where the variation in x_t, from realization to realization, due to a jump in $F(\lambda)$ at the origin, has a clearly distinguished physical meaning (corresponding to a zeroing error, for example, in some mechanism, which might vary from run to run). However, in situations where only one realization (or a part of one) is ever available the part of x_t, due to the jump in $F(\lambda)$ at the origin, will be absorbed into the mean so that for the residual process, without a jump at the origin in its spectral distribution function, the sample mean will be a consistent estimator of the true mean.

In any case

$$\lim_{n \to \infty} \mathscr{E}\{\bar{x}^2\} = \lim_{n \to \infty} \frac{1}{r} \sum_{0}^{n-1} \gamma_t$$

since

$$\frac{1}{n} \sum_{0}^{n-1} \gamma = \int_{-\pi}^{\pi} \frac{1}{n} \sum_{0}^{n-1} e^{it\lambda} \, dF(\lambda) \longrightarrow dF(0)$$

again by bounded convergence. Thus $\bar{x}$ converges in the mean to zero if and only if $n^{-1} \sum_{0}^{n-1} \gamma_t$ converges to zero.

[1] That is $\mathscr{E}\{| \bar{x} - (z(0) - z(0^2 -)) |\} \longrightarrow 0$.

If we are willing to impose further restrictions on the process $\{x_t\}$ we can obtain an analogous result relating to the serial covariances γ. Let us assume that for the stationary process x_t, and each fixed t,

(1) $\mathscr{E}\{(x_{s+t}x_s)^2\} < \infty \quad s = 0, \pm 1, \ldots$

(2) $\mathscr{E}\{x_{s_1+s_2+t}x_{s_1+s_2}x_{s_1+t}x_{s_1}\}$ is independent of s_1.

Both of these conditions would be true, of course, if the process were strictly stationary and the expectations existed. If the conditions hold the process, $y_s = x_{s+t}x_s - \gamma_t$, is stationary and we are led to consider the convergence of

$$\frac{1}{n}\sum_{1}^{n} y_s.$$

We are now, with respect to $\{y_s\}$, in a completely analogous position to that with respect to the initial $\{x_t\}$ and thus can state that

$$\frac{1}{n}\sum_{1}^{n} x_{s+t}x_s$$

converges in the mean to γ_t if and only if $n^{-1}\sum_{s=0}^{n-1} \mathscr{E}\{y_0 y_s\}$ converges to zero. This is equivalent to the condition

$$\frac{1}{n}\sum_{s=0}^{n-1} \mathscr{E}\{x_{s+t}x_s x_t x_0\} \to \gamma_t^2. \tag{1}$$

Thus $c_t' = n^{-1}\sum_{s=0}^{n-1} x_{s+t}x_s$ is a consistent estimate of γ_t if and only if this last condition is satisfied. (The reason for the prime will be apparent later.) If $x_s x_{s+t}$ and $x_0 x_t$ were independent for $s > s_0$ then

$$\mathscr{E}\{x_s x_{s+t} x_t x_0\} = \gamma_t^2, \quad s > s_0$$

and the sum would indeed converge to γ_t^2. Thus the condition for the consistency of c_t' may be interpreted as a condition on the rate at which the influence of x_0 on x_s falls off as $|s|$ is increased.

Example

Let us consider a process of the form

$$x_t = \sum_0^\infty \alpha_j \varepsilon_{t-j}, \quad \sum_0^\infty \alpha_j^2 < \infty$$

where the ε_t are independent with zero mean and unit variance. The requirement of unit variance is no restriction since the scale constant may be absorbed into the α_j. Such a process is called a linear process. We further require that the fourth cumulant of ε_t, κ_4, should be finite.

Then we easily derive

$$\begin{aligned}\mathscr{E}\{x_s x_{s+t} x_t x_0\} &= \gamma_t^2 + \gamma_s^2 + \gamma_{t+s}\gamma_{t-s} + \kappa_4 \sum_{j=0}^{\infty} \alpha_j \alpha_{j+s} \alpha_{j+t} \alpha_{j+s+t} \\ &= \gamma_t^2 + \gamma_s^2 + \gamma_{s+t}\gamma_{s-t} + K(s, t),\end{aligned}$$

let us say, where $\alpha_j = 0$ for $j < 0$.

The condition (1) thus becomes

$$\frac{1}{n}\sum_{s=0}^{n-1} \gamma_s^2 + \frac{1}{n}\sum_{s=0}^{n-1} \gamma_{s+t}\gamma_{s-t} + \frac{1}{n}\sum_{s=0}^{n-1} K(s, t) \to 0.$$

A necessary and sufficient condition for the first two terms to tend to zero, for every t, is evidently

$$\frac{1}{n}\sum_{s=0}^{n-1} \gamma_s^2 \to 0. \tag{2}$$

If we put

$$\sum_0^\infty \alpha_j e^{ij\lambda} = h(\lambda)$$

and
$$k_t(\lambda) = h(\lambda) * h(\lambda) e^{-it\lambda}$$

where by the star we here indicate the formation of the convolution of the two functions (which exists and is continuous and square integrable since $h(\lambda)$ is square integrable),[1] then $K(s, t)$ is the sth

[1] Littlewood [1] p. 29 Theorem 7.

Fourier coefficient of $K_t(\lambda) = \kappa_4 k_t(\lambda) k_t(-\lambda)$ (which is integrable since $k_t(\lambda)$ is square integrable). Thus

$$\frac{1}{n}\sum_{s=0}^{n-1} K(s, t) = \int_{-\pi}^{\pi} K_t(\lambda)\frac{1}{n}\sum_{0}^{n-1} e^{is\lambda}\, d\lambda$$

and $\frac{1}{n}\sum_{0}^{n-1} e^{is\lambda}$ is bounded in absolute value by unity and converges to a function which is zero everywhere save at $\lambda = 0$ where it is unity. Thus, by dominated convergence, the sum converges to zero.

Thus the necessary and sufficient condition for the mean convergence of c_t' to γ_t for all t is the condition (2) above. However

$$\frac{1}{n}\sum_{0}^{n-1} \gamma_t^2 = \int_{-\pi}^{\pi}\int_{-\pi}^{\pi} \frac{1}{n}\sum_{0}^{n-1} e^{it(\lambda_1-\lambda_2)} f(\lambda_1) f(\lambda_2)\, d\lambda_1\, d\lambda_2$$

and by a repetition of the type of argument just given this is seen to converge to zero.

We shall investigate the variance of the consistent statistics c_t' (or slight modifications of them) in the next section.

3. The correlogram

So far, in our theoretical discussions, we have represented the Fourier transform of $f(\lambda)$ in terms of the γ_t. From the point of view of the structure of the series the scale of measurement of the observations is irrelevant and we prefer to work in terms of the serial correlations

$$\rho_t = \gamma_t/\gamma_0.$$

It will be seen presently that these are also preferable from a statistical point of view. The set of ρ_t; $t = 0, 1, \ldots$: is called the correlogram. Needless to say it specifies the process to the same extent as its Fourier–Stieltjes transform does. Whether we prefer to think in terms of the spectrum or the correlogram depends upon a number of considerations amongst which the physical context is of prime importance.

In many applications the spectrum has a very real physical meaning. As an example one might consider the motion of the waves at a jetty. Here the x_t would be the height of the waves above some base point at equidistant intervals of time (clearly a case where the underlying process is one in continuous time). Frequencies of oscillation in certain, perhaps quite narrow, ranges might then be of interest since they might correspond to some natural frequencies of ships which might be caused by such oscillations to bump heavily against the jetty. Numerous other examples occur, particularly in electrical engineering.

On the other hand the autoregressive model (see I.4) has been of considerable importance in econometric work. If one considers a stock market, for example, one could expect that buyers and sellers today in determining their offers would be affected by the previous day's price and by its change from the day before. If these two relations had an additive effect and x_t is today's price for a stock then one expects a relation of the form

$$x_t = \alpha_1 x_{t-1} + \alpha_2(x_{t-1} - x_{t-2}).$$

In addition, however, there would usually be some new information (which can be taken to be independent of x_{t-1} and x_{t-2} since otherwise it would have been reflected therein and would not be new) which would lead to a relation of the form

$$x_t + \beta_1 x_{t-1} + \beta_2 x_{t-2} = \varepsilon_t.$$

This analysis is clearly grossly over-simple. However, if one has a small amount of information available one has, beforehand, to restrict oneself to a small range of models and the autoregressive model appears to be the most appropriate in many economic contexts. As we shall see the constants β_j, $(j = 1, \ldots, p)$ in an autoregression are estimated by estimating first the ρ_t $(t = 1, \ldots, p)$. In this situation one may prefer to think in terms of the correlogram rather than the spectrum and a consideration of it would be a reasonable method of proceeding to discover, for example, a suitable maximum lag, p, in the model.

Another factor, closely related to the physical context, is the final purpose of the analysis. If, as in the example before the last, the ultimate purpose is the investigation of the importance of certain ranges of frequencies, then of course we need to know the spectral function in these ranges. In other cases, however, we may be concerned rather to find a simple transformation which will reduce the process to something approaching a process of independent random variables, so that a later stage of statistical analysis may be proceeded with. Here again the autoregressive process and the consequent consideration of the ρ_t could be of prime importance. After all, if one is fundamentally concerned with eliminating the relations between the x_t it is natural to study the ρ which measure the intensity of these relations.

A third factor, whose importance will become evident in what follows, is mathematical convenience. From one point of view the correlogram is simpler since the set (the integers) on which it is defined is discrete. From another point of view the spectral function is simpler since it is defined on a compact set.

The extent to which one can interpret the fine structure of a correlogram seems limited and it appears that one must concentrate on certain features such as pronounced oscillations and the speed with which the ρ_t converge to zero. At the same time it is necessary to be able to correlate such observations with some more directly interpretable model of the stochastic process; that is, one needs to have some canonical types of process the structures of whose correlograms can be compared with that of the process being considered. We go on to consider the behaviour of the correlogram of the examples given in I.3.[1]

(1) $F(\lambda) = F_1(\lambda); f(\lambda) = \gamma_0/(2\pi)$. Then $\rho_t = 0,\ t \neq 0$.

(2) $F(\lambda) = F_2(\lambda)$ with p jumps at the points λ_j, the jth jump being f_j. Here $\rho_t = \gamma_0^{-1} \int_{-\pi}^{\pi} e^{it\lambda}\, dF(\lambda) = \gamma_0^{-1} \sum_j f_j e^{it\lambda_j}$ so that the correlogram is almost periodic being a sum of oscillations with frequencies λ_j.

A process whose $F(\lambda)$ is of the form $F_1(\lambda) + F_2(\lambda)$ with two parts as in these first two examples will have a correlogram of the same form (except for ρ_0).[2] However, the component $F_1(\lambda)$ will introduce

[1] A more detailed discussion of the nature of the correlogram of processes of these types may be found in Kendall [2] Chapters 29 and 30.

[2] Of course $\rho_0 = 1$, but for the example being considered the maxima of ρ_t, for $t \neq 0$, will be less than unity.

an essential difference when estimation of the correlogram is considered.

(3) $$F(\lambda) = F_1(\lambda);\ f(\lambda) = \frac{1}{2\pi}\sum_{-q}^{q} \gamma_t e^{it\lambda}.$$

Here the correlogram is characterized by the fact that $\rho_t = 0,\ t > p$.

(4) $F(\lambda) = F_1(\lambda);\quad f(\lambda) = K\left|\sum_0^p \beta_j e^{ij\lambda}\right|^{-2}$ where all roots of $\sum_0^p \beta_j z^j$ lie outside of the unit circle. (The choice of this polynomial $\sum_0^p \beta_j e^{ij\lambda}$ instead of one of its $2^p - 1$ counterparts is of course irrelevant insofar as the behaviour of the correlogram is concerned.) From the relation $\sum_0^p \beta_j x_{t-j} = \varepsilon_t$ with ε_t orthogonal to x_{t-j} for $j > 0$ we see that the ρ_t satisfy

$$\sum_0^p \beta_j \rho_{t-j} = 0;\quad t > 0$$

so that $\rho_t = \sum_j a_j z_j^t,\ t \geqslant 0$ where the z_j are the roots of $\sum_0^p \beta_j z^{p-j}$ while the a_j are determined from the fact that $\rho_t = \rho_{-t}$. If any z_j are complex, trigonometric terms will of course appear in the formula. Since $|z_j| < 1$ the correlogram will decay exponentially with possibly damped oscillations.

In case $p = 1$ we have $\rho_t = \rho^{|t|}$ where $\rho = -\beta_0^{-1}\beta_1$.

For $p = 2$ the correlogram will be a damped sine curve if $\beta_0\beta_2 > \frac{1}{4}\beta_1^2$.

(5) $$F(\lambda) = F_1(\lambda);\quad f(\lambda) = K\left|\sum_0^q \alpha_j e^{ij\lambda}\right|^2 \Big/ \left|\sum_0^p \beta_j e^{ij\lambda}\right|^2.$$

If we consider the representation

$$\sum_0^p \beta_j x_{t-j} = \sum_0^q \alpha_j \varepsilon_{t-j}$$

where all roots of $\sum_0^p \beta_j z^{p-j}$ lie inside of the unit circle, we see that the ρ_t satisfy

$$\sum_0^p \beta_j \rho_{t-j} = 0; \quad t > q.$$

The first q of the ρ_t of course depend upon the α_j. Thus the ρ_t here behave ultimately as in example (4) above save for the first q.

A process of this form will arise when we observe a process of the type discussed under (4) but with a superimposed error of observation, η_t say, which is orthogonal to ε_t and for which $\mathscr{E}\{\eta_s\eta_t\} = 0$, $t \neq s$. The spectral density will then be $K_1 \left| \sum_0^p \beta_j e^{ij\lambda} \right|^{-2} + K_2$ which is clearly of the present form.

So far we have discussed the theoretical nature of the correlogram. We now consider its estimation. Since n, the sample size, will be fixed it is now preferable to adopt the notation

$$c_t = (n-t)^{-1} \sum_1^{n-t} x_{s+t} x_s$$

for the estimator of γ_t. We have already, to some extent, justified the use of an estimator of this form and we shall justify it further later. Let us consider a linear process where the ε_t have finite fourth cumulant κ_4. We have, using the results and notation of the example of the last section, writing var (x) for the variance of x, and noting that $\mathscr{E}(c_t) = \gamma_t$,

$$\operatorname{var}(c_t) = \mathscr{E}(c_t^2) - \gamma_t^2$$

$$= \left(\frac{1}{n-t}\right)^2 \sum_{\substack{i,j\\=1}}^{n-t}\sum \{\gamma_{i-j}^2 + \gamma_{i-j-t}\gamma_{i-j+t} + K(i-j,t)\}$$

$$= \frac{1}{n-t} \sum_{s=-(n-t)+1}^{(n-t)-1} \left(1 - \frac{|s|}{n-t}\right)\{\gamma_s^2 + \gamma_{s-t}\gamma_{s+t} + K(s,t)\}. \quad (1)$$

Thus

$$(n-t)\operatorname{var}(c_t) \longrightarrow 2\pi \int_{-\pi}^{\pi} f^2(\lambda)(1 + e^{i2t\lambda})\, d\lambda + \kappa_4\left\{\int_{-\pi}^{\pi} e^{it\lambda} f(\lambda)\, d\lambda\right\}^2 \quad (2)$$

since, for example,

$$\sum_{-(n-t)+1}^{n-t-} \left(1 - \frac{|s|}{n-t}\right) K(s,t)$$

is a Caesaro sum of the Fourier series for the continuous function $K_t(\lambda)$, at $\lambda = 0$. The right-hand side of (2) may of course be infinite whereas (1) always converges to zero, as we have seen in the last section. Nevertheless the case where $f(\lambda)$ is square integrable so that the variance of c_t is of order n^{-1} is probably the most important and we shall concentrate upon it in what follows.

In a similar fashion one can show that (writing 'cov' for covariance)

$$\operatorname{cov}(c_t, c_{t+u}) = \frac{1}{n-t}\frac{1}{n-t-u} \sum_{j=1}^{n-t-u} \sum_{i=1}^{n-t} \{\gamma_{i-j-u}\gamma_{i-j} + \gamma_{i-j-t-u}\gamma_{i-j+t} + K(i-j,t,u)\} \quad (3)$$

where

$$K(s,t,u) = \kappa_4 \sum_0^{\infty} \alpha_j \alpha_{j+s} \alpha_{j+t} \alpha_{j+s+t+u}.$$

Thus

$$n \operatorname{cov}(c_t, c_{t+u}) \longrightarrow 2\pi \int_{-\pi}^{\pi} f^2(\lambda)\{e^{iu\lambda} + e^{i(2t+u)\lambda}\}\, d\lambda$$

$$+ \kappa_4 \int_{-\pi}^{\pi} e^{it\lambda} f(\lambda)\, d\lambda \int_{-\pi}^{\pi} e^{i(t+u)} f(\lambda)\, d\lambda$$

$$= \sum_{-\infty}^{\infty} (\gamma_s \gamma_{s+u} + \gamma_{s+t+u}\gamma_{s-t}) + \kappa_4 \gamma_t \gamma_{t+u} \tag{4}$$

when $f(\lambda)$ is square integrable. The formula is due to Bartlett [2]. When the eighth moment of ε_t is finite and $f(\lambda)$ is square integrable the same method shows that $\mathscr{E}\{(c_t - \gamma_t)^4\}$ is of order n^{-2}.

Under these conditions we may, neglecting terms of higher order than n^{-1}, obtain the covariances of the statistics

$$r_t = c_t/c_0$$

by replacing the r_t by the first two terms in their Taylor series expressions, as a function of c_t and c_0, about γ_t and γ_0 (see Lomnicki and Zaremba [1]). Thus

$$\lim_n n \operatorname{cov}(r_t, r_{t+u})$$

$$= \frac{\operatorname{cov}(c_t, c_{t+u})}{\gamma_0^2} - \frac{\gamma_t \operatorname{cov}(c_0, c_{t+u})}{\gamma_0^3}$$

$$- \frac{\gamma_{t+u} \operatorname{cov}(c_0, c_t)}{\gamma_0^3} + \frac{\gamma_t \gamma_{t+u} \operatorname{var}(c_0)}{\gamma_0^4}$$

$$= \sum_{-\infty}^{\infty} \{\rho_s \rho_{s+u} + \rho_{s+t+u}\rho_{s-t} - 2\rho_{t+u}\rho_s\rho_{s-t} - 2\rho_t\rho_s\rho_{s-t-u} + 2\rho_t\rho_{t+u}\rho_s^2\} \tag{5}$$

the term involving κ_4 *disappearing*. These formulae are again due to Bartlett [2].

In a similar fashion we obtain

$$n\,\mathscr{E}\{(r_t - \rho_t)^2\} = 2 \sum_{-\infty}^{\infty} (\rho_s^2 \rho_t - \rho_s \rho_{s-t}). \tag{6}$$

The ρ_t will decrease to zero as t increases (Riemann–Lebesgue lemma). However, the variance of the r_t will, for fixed n, not do so, of course. Thus the coefficient of variation of the r_t will tend to increase as t increases and the later r_t will provide practically no information. In practice one will have to make a decision as to where the correlogram should be truncated and will study the appearance of the remainder of the correlogram in order to 'type' the process. We will study this problem of truncation of the correlogram, in another form, below.

No very general theorems concerning the asymptotic normality of the r_t appear to be available. The problem can be reduced to that of the joint asymptotic normality of c_t and c_0 since if the distribution of the $\sqrt{n}(c_t - \gamma_t)$ tends to the normal distribution that of $\sqrt{n}(r_t - \rho_t)$ will tend to normality also. From the theorem quoted in the first section of this chapter it is evident that the distribution of $\sqrt{n}(c_t - \gamma_t)$ will tend to normality if x_t is strictly stationary with finite fourth moment and m-dependent since $x_s x_{s+t}$ will then be strictly stationary and $(m + t)$ dependent with finite variance. In case $m = 0$ (independence) the condition on the fourth moment may be dropped and we need require only a finite variance for x_t for then $x_s x_{s+t}$ is t-dependent and has finite variance without any assumption being made about the fourth moment of x_t. Thus $\sqrt{n}\, r_1$, which as we shall see is a suitable statistic for testing the lack of serial dependence in a process, is asymptotically normal under very general conditions.

The condition of finite dependence, for the asymptotic normality of the $\sqrt{n}(r_t - \rho_t)$, is much too strict and can be relaxed considerably. One will clearly have to impose some restriction which will make the α_j fall off to zero sufficiently fast and we shall merely mention here that the condition $\sum_0^\infty |j\alpha_j| < \infty$ is sufficient (Walker [1]).

Example 1

In fig. 1 the first 50 values of the observed sample serial correlations of a second order autoregression

$$x_t - 1{\cdot}1x_{t-1} + 0{\cdot}5x_{t-2} = \varepsilon_t$$

are shown, the observed correlogram having been computed from

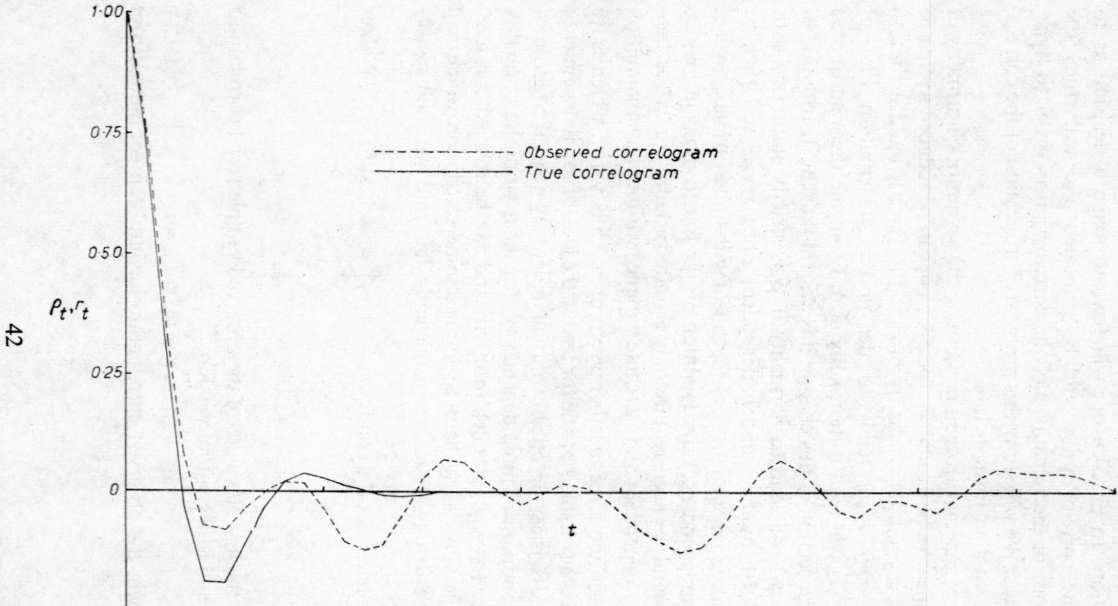

Fig. 1. Correlogram of constructed autoregressive series of example 1, from 480 observations, compared with true correlogram

480 observations.[1] The formulae used for the estimates, r_t, of ρ_t differed from those given above. The denominator was replaced by

$$\frac{1}{n-t}\left(\sum_{1}^{n-t} x_s^2 \sum_{1}^{n-t} x_{s+t}^2\right)^{1/2}$$

and mean corrections were made (a question which will be discussed in Chapter V). The difference introduced by these changes in definition will hardly be perceptible in a sample of 480 observations.

The failure of the sample correlogram to damp down is very evident. From (5) above we see that the variance of the r_t approaches, in the present circumstances,

$$n^{-1}\sum_{-\infty}^{\infty} \rho_s^2$$

as s increases, which in the present case is approximately 0·005. Thus a two standard deviations range about the horizontal axis will cover values (roughly) between $\pm 0{\cdot}14$. This limit is not exceeded in the 30 plotted values between $t = 20$ and $t = 50$. Adjacent r_t are of course correlated (see (5)), which explains the relatively smooth appearance of the sample correlogram. It is evident that in this case it would be difficult to distinguish between a moving average and an autoregression on the basis of the appearance of the correlogram since the oscillations damp down so quickly that they are swamped by the sampling fluctuations.

Example 2

The r_t shown in Table 1 are the serial correlations of the logarithms[2] of the annual trappings of Canadian Lynx over the years 1821 to 1934 (inclusive) (see Moran [1]). They are graphed in fig. 2. The oscillations persist in a most remarkable manner. Such an oscillation could of course be obtained from an autoregression. For example

[1] These are taken from Kendall [1] Chapter 3.

[2] Logarithms were taken to make the distribution of the x_t more nearly normal. Again the r_t were formed after mean corrections were made.

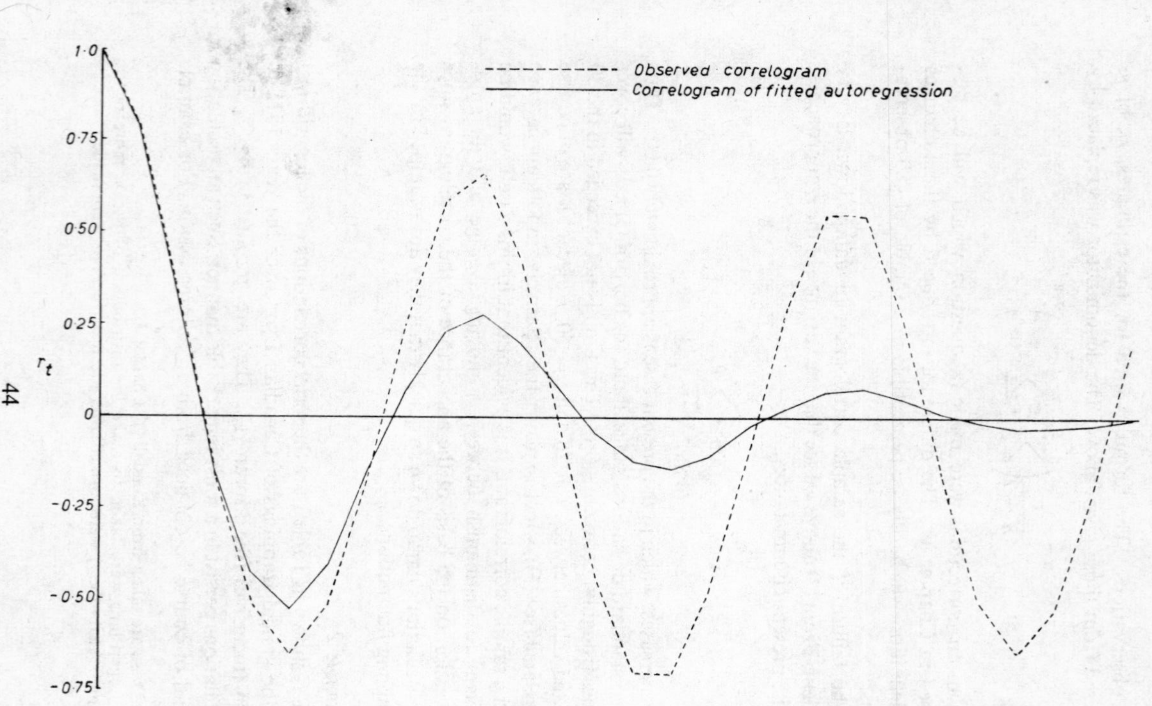

Fig. 2. Correlogram of Lynx data, from 114 observations, and correlogram of process generated by $x_t - 1{\cdot}41x_{t-1} + 0{\cdot}077x_{t-2} = \epsilon_t$

TABLE 1

t	r_t	t	r_t	t	r_t	t	r_t
1	0·79516	8	0·25296	15	−0·70538	22	−0·14559
2	0·34788	9	0·58422	16	−0·47523	23	−0·50129
3	−0·13596	10	0·66419	17	−0·08560	24	−0·64342
4	−0·51332	11	0·42315	18	0·30214	25	−0·53101
5	−0·65116	12	−0·01529	19	0·54795	26	−0·21046
6	−0·51696	13	−0·43674	20	0·54308	27	0·19955
7	−0·16895	14	−0·69531	21	0·26726		

the relation (see below for a further discussion)

$$x_t - 1{\cdot}41x_{t-1} + 0{\cdot}77x_{t-2} = \varepsilon_t$$

will generate a process having a correlogram given by

$$\rho_t = (0{\cdot}88)^t \frac{\sin(0{\cdot}64t + 1{\cdot}74)}{\sin(1{\cdot}74)}.$$

This is also graphed in fig. 2. The fit is evidently not very good.

4. Estimating the parameters of finite parameter schemes

In Chapter I we showed that (neglecting the contribution from the singular part, $F_3(\lambda)$, of the spectrum) any non-deterministic stationary process could be represented in the form

$$x_t = \sum_{-\infty}^{\infty} z_j e^{it\lambda_j} + \sum_{0}^{\infty} \alpha_j \varepsilon_{t-j}. \tag{1}$$

In this section we shall neglect the contribution from the first term in (1) and shall concentrate on processes having an absolutely continuous spectral function. We shall return to the consideration of the jumps in the spectral function in the next chapter and in the last.

In any case, of course, one cannot estimate an infinite set of parameters given only a sequence of n consecutive x_t. Thus one must be ready to impose, *a priori*, some restriction on the nature of the process. These prior restrictions may be imposed in different ways. In the following chapter we shall impose *a priori* restrictions by means of assumptions concerning the smoothness of the function $f(\lambda)$.

Here we shall instead consider cases where the α_j are functions of a relatively small number of parameters. This would be so, for example, if x_t were generated by an autoregression.

One can invest the problem with a considerable degree of generality by considering a process of the form $\sum_0^\infty \alpha_j \varepsilon_{t-j}$ where the ε_t are *independent* with finite fourth cumulant and for which the spectral density

$$f(\lambda) = \frac{\sigma^2}{2\pi} \left| \sum_0^\infty \alpha_j e^{ij\lambda} \right|^2$$

is defined by q parameters $\theta_1, \ldots, \theta_q$, in addition to σ^2 (so that each α_j is a specified function of these parameters). The method of least squares then consists in minimizing, with respect to the θ_j, the quadratic form

$$Q_n(\theta_1, \ldots, \theta_q) = \sigma^2 n^{-1} \mathbf{x}' \mathbf{\Gamma}_n^{-1} \mathbf{x}$$

where $\mathbf{x}$ is the vector of n observed values of the process while $\mathbf{\Gamma}_n$ was defined by (I.5.6). Evidently Q_n is a function only of the indicated parameters. The parameter σ^2 will be estimated by the minimized value of Q_n. The derivation of the asymptotic properties of the estimates, under specified conditions on the regularity of the α_j as functions of the parameters and upon the rate of approach to zero of the α_j, is by no means simple. Whittle [2] has considered the problem in this generality and has shown[1] that

(1) The least squares equations have one system of solutions which converges to the true set of values of the θ_j in probability (in general there will be more than one solution).

(2) The covariance matrix of the estimates $\hat{\theta}_j$ is given by

$$\frac{2}{n}\left[\frac{1}{2\pi}\int_{-\pi}^{\pi} \frac{\partial \log 2\pi\sigma^2 f(\lambda)}{\partial \theta_i} \frac{\partial \log 2\pi\sigma^2 f(\lambda)}{\partial \theta_j} d\lambda\right]^{-1}$$

[1] Whittle's extremely interesting and original arguments are by no means rigorous but there seems little doubt that his results hold true under reasonable conditions.

where the element in the ith row and jth column of the matrix is shown.

(3) $\hat{\sigma}^2$ is uncorrelated with the $\hat{\theta}_j$.

The $\hat{\theta}_j$ will be functions of the c_t save for 'end effects' whose influence will vanish asymptotically. The influence of the c_t on $\hat{\theta}_j$ will decrease, as n increases, at a rate dependent upon that at which the α_j go to zero. It can be expected therefore that for sufficiently rapid rates of approach of these α_j to zero the $\hat{\theta}_j$ will be asymptotically normal.

In practice, however, the estimation of the constants in a scheme specified in this way is extremely troublesome. For example, when

$$x_t = \varepsilon_t + \alpha\varepsilon_{t-1}$$

then

$$\frac{1}{\sigma^2}\mathbf{\Gamma}_n = \begin{bmatrix} 1+\alpha^2 & \alpha & 0 & \cdots & 0 \\ \alpha & 1+\alpha^2 & \alpha & \cdots & 0 \\ 0 & \alpha & 1+\alpha^2 & \cdots & 0 \\ \cdot & \cdot & \cdot & & \cdot \\ \cdot & \cdot & \cdot & & \cdot \\ \cdot & \cdot & \cdot & & \cdot \\ 0 & 0 & 0 & \cdots & 1+\alpha^2 \end{bmatrix}$$

whose inverse is close[1] to

$$\frac{1}{1-\alpha^2}\begin{bmatrix} 1 & -\alpha & \alpha^2 & \cdots & (-\alpha)^{n-1} \\ -\alpha & 1 & -\alpha & \cdots & (-\alpha)^{n-2} \\ \alpha^2 & -\alpha & 1 & \cdots & (-\alpha)^{n-3} \\ \cdot & \cdot & \cdot & & \cdot \\ \cdot & \cdot & \cdot & & \cdot \\ \cdot & \cdot & \cdot & & \cdot \\ (-\alpha)^{n-1} & (-\alpha)^{n-2} & (-\alpha)^{n-3} & \cdots & 1 \end{bmatrix}$$

so that Q_n becomes, approximately,

$$\frac{1}{1-\alpha^2}\sum_{-n+1}^{n-1} c_t\left(1-\frac{|t|}{n}\right)(-\alpha)^t. \tag{2}$$

The solution of the equation obtained by putting the derivative of

[1] The true inverse is more complicated.

this equal to zero will clearly be difficult and will of course involve a problem of identifying the appropriate root as that giving the estimator of α.

If the terms up to $t = 1$, only, are used the resulting estimate is obtained from

$$\alpha c_0 - (n - 1)c_1(1 + \alpha^2)/n = 0$$

or, approximately,

$$r_1 = \alpha/(1 + \alpha^2).$$

This gives $\tilde{\alpha} = \frac{1}{2}\{r_1^{-1} \pm \sqrt{r_1^{-2} - 1}\}$, one of which must be chosen on prior grounds.

However, the neglect of the remaining terms in (2) will involve a considerable loss of information unless α is near to zero. Indeed the asymptotic efficiency of this estimate may be shown to be (Whittle [2])

$$(1 - \alpha^2)^3/(1 + \alpha^2 + 4\alpha^4 + \alpha^6 + \alpha^8)$$

which is only 0·27 for $\alpha = \frac{1}{2}$ and approaches zero as α approaches unity.

Partly because of these difficulties the finite parameter models used in practice have nearly always been autoregressions, for which, we shall see, the estimation equations are much simpler. This is, of course, not the only reason for the use of autoregressive models. They have an obvious appeal, in many applications, on prior grounds.

Let us consider therefore the estimation, by least squares, of the parameters of a general autoregressive scheme,

$$\sum_0^p \beta_j x_{t-j} = \varepsilon_t, \quad \beta_0 = 1.$$

As we have seen in Chapter I some restriction must be placed on the β_j in order to uniquely identify them. The restriction we shall make is that the roots of $\sum_0^p \beta_j z^{p-j}$ should be inside the unit circle. This amounts to specifying that x_t should be a moving average of the

ε_{t-j}, $j \geqslant 0$ (with weights which tend to zero exponentially).[1] We also assume that the fourth moment of the ε_t is finite and that they are independent, and shall put var $(\varepsilon_t) = \sigma^2$. The estimation equations now become (after slight modification to eliminate end effects whose order will be n^{-1}).

$$\sum_{j=0}^{p} \hat{\beta}_j \left\{ \sum_{t=p+1}^{n} x_{t-j} x_{t-k} \right\} = 0, \quad k = 1, \ldots, p \tag{3}$$

$$= \sum_{j=0}^{p} \hat{\beta}_j (n-p) c_{j,k}, \quad \text{let us say.}$$

Following Mann and Wald [1] we introduce the quantities

$$y_k = (n-p)^{-\frac{1}{2}} \sum_{t=p+1}^{n} \left(\sum_{0}^{p} \beta_j x_{t-j} \right) x_{t-k} = (n-p)^{-\frac{1}{2}} \sum_{t=p+1}^{n} \varepsilon_t x_{t-k}$$

$$= \sum_{0}^{p} \beta_j (n-p)^{\frac{1}{2}} c_{j,k}, \quad k = 1, \ldots, p. \tag{4}$$

Then from (3) we have

$$-y_k = (n-p)^{\frac{1}{2}} \sum_{j=1}^{p} (\hat{\beta}_j - \beta_j) c_{j,k}.$$

This may be written in the form

$$\mathbf{C}_p (n-p)^{\frac{1}{2}} (\hat{\boldsymbol{\beta}} - \boldsymbol{\beta}) = -\mathbf{y}$$

where $\mathbf{C}_p$ is a $(p \times p)$ matrix which converges in probability to the (non-singular) matrix $\boldsymbol{\Gamma}_p$ while the vectors $(\hat{\boldsymbol{\beta}} - \boldsymbol{\beta})$ and $\mathbf{y}$ have $\hat{\beta}_j - \beta_j$ and y_j in the jth place. Moreover, the elements of the vector $\mathbf{y}$ are asymptotically jointly normally distributed with zero means and covariance matrix $\sigma^2 \boldsymbol{\Gamma}_p$. The asymptotic normality follows from the

[1] If one is specifying an autoregressive process as that generating the observations one will certainly expect that the process will be such that only past and present values of ϵ_t affect x_t so that this condition appears to be no real additional restriction.

third expression on the right side of (4) since

$$(n-p)^{\frac{1}{2}}c_{j,k} = (n-p)^{\frac{1}{2}}c_{j-k} + 0(n^{-\frac{1}{2}})$$

and $(n-p)^{\frac{1}{2}}c_{j-k}$ is asymptotically normal. (The requirements of the result due to Walker quoted before example 1 of section 3 are clearly satisfied here.) That the covariance matrix is $\sigma^2\mathbf{\Gamma}_p$ is easily checked.

Thus $(n-p)^{\frac{1}{2}}(\hat{\boldsymbol{\beta}} - \boldsymbol{\beta})$ converges in probability to $\mathbf{\Gamma}_p^{-1}\mathbf{y}$ and is thus asymptotically normally distributed with zero mean and covariance matrix

$$\mathscr{E}\{\mathbf{\Gamma}_p^{-1}\mathbf{y}'\mathbf{y}\mathbf{\Gamma}_p^{-1}\} = \sigma^2\mathbf{\Gamma}_p^{-1}.$$

In practice the $c_{j,k}$ would be replaced by c_{j-k}. We shall indicate the corresponding matrix by $\mathbf{G}_p$. It follows that

$$\hat{\sigma}^2 \equiv (n-p)^{-1}\sum_{t=p+1}^{n}\left\{\sum_{j=0}^{p}\hat{\beta}_j x_{t-j}\right\}^2$$

converges in probability to σ^2.

Example 1

We consider once more the first example of section 3. The first two serial correlations of this series are $r_1 = 0{\cdot}762$ and $r_2 = 0{\cdot}377$. The equations (3) are thus equivalent to

$$0{\cdot}762 + \hat{\beta}_1 + 0{\cdot}762\hat{\beta}_2 = 0$$
$$0{\cdot}377 + 0{\cdot}762\hat{\beta}_1 + \hat{\beta}_2 = 0$$

which give the estimates

$$\hat{\beta}_1 = -\ 1{\cdot}132, \quad \hat{\beta}_2 = 0{\cdot}485.$$

We may estimate $(n-p)^{-1}\sigma^2\mathbf{\Gamma}_p^{-1}$ by

$$(n-p)^{-1}\hat{\sigma}^2\mathbf{G}_p^{-1} = \begin{bmatrix} 0{\cdot}0016 & -0{\cdot}0012 \\ -0{\cdot}0012 & 0{\cdot}0016 \end{bmatrix}.$$

One could now determine an approximate confidence region for β_1 and β_2. It is clear that the true values will be well within a region corresponding to any reasonable confidence coefficient since both estimates differ from the true value by less than one standard deviation.

Example 2

We again consider example 2 of section 3. Here the equations have the solution

$$\hat{\beta}_1 = -1{\cdot}41, \quad \hat{\beta}_2 = 0{\cdot}77$$

while

$$(n-p)^{-1}\hat{\sigma}^2\mathbf{G}_p^{-1} = \begin{bmatrix} 0{\cdot}068 & -0{\cdot}054 \\ -0{\cdot}054 & 0{\cdot}068 \end{bmatrix}.$$

An approximate 5% confidence region is given by

$$39{\cdot}8\{(\beta_1 + 1{\cdot}41)^2 + (\beta_2 - 0{\cdot}77)^2\} + 63{\cdot}2\{(\beta_1 + 1{\cdot}41)(\beta_2 - 0{\cdot}77)\} \leqslant 5{\cdot}99.$$

CHAPTER III

Estimation of the Spectral Density and Distribution Functions

1. The periodogram

If one wishes to test for a jump in the spectral function at λ one is led, from standard regression theory, to introduce the statistic (the periodogram),

$$I_n(\lambda) = \frac{2}{n}\left|\sum_{1}^{n} x_t e^{it\lambda}\right|^2. \tag{1}$$

The introduction of the factor 2 is conventional (and as we shall see below convenient).

If we rearrange the double sum involved in (1) we see that

$$I_n(\lambda) = 2\sum_{-n+1}^{n-1}\left(1-\frac{|t|}{n}\right)c_t e^{it\lambda}$$

which has the expectation

$$2\sum_{-n+1}^{n-1}\left(1-\frac{|t|}{n}\right)\gamma_t e^{it\lambda} \to 4\pi f(\lambda), \quad \text{a.e.} \tag{2}$$

In the next chapter we shall return to the problem of testing for jumps in $F(\lambda)$. *For the present chapter we shall restrict ourselves to the case of absolutely continuous* $F(\lambda)$, and shall consider the problem of estimating $f(\lambda)$. We see from (2) that $I_n(\lambda)/(4\pi)$ is an asymptotically unbiased estimator of this function and we are led to consider its covariance properties. Let us consider first the case where the process is one of independent random variables ε_t with fourth cumulant κ_4 and variance σ^2. Then

$$\mathscr{E}\{I_n(\lambda_1)I_n(\lambda_2)\} = \mathscr{E}\left[\frac{4}{n^2}\sum_v\sum_u\sum_t\sum_s \varepsilon_s\varepsilon_t\varepsilon_u\varepsilon_v e^{i\{(s-t)\lambda_1+(u-v)\lambda_2\}}\right]$$

$$= \frac{4}{n^2}\left[\sigma^4\left\{n^2 + \sum_t\sum_s e^{i(s-t)(\lambda_1+\lambda_2)} + \sum_t\sum_s e^{i(s-t)(\lambda_1-\lambda_2)}\right\} + n\kappa_4\right]$$

$$= \frac{4\kappa_4}{n} + 4\sigma^4 + \frac{4\sigma^4}{n^2}\left\{\left(\frac{\sin\frac{1}{2}n(\lambda_1+\lambda_2)}{\sin\frac{1}{2}(\lambda_1+\lambda_2)}\right)^2 + \left(\frac{\sin\frac{1}{2}n(\lambda_1-\lambda_2)}{\sin\frac{1}{2}(\lambda_1-\lambda_2)}\right)^2\right\}.$$

Thus

$$\left.\begin{aligned}
\text{var}\,\{I_n(\lambda)\} &= 4\sigma^4 + \frac{4\kappa_4}{n} + \frac{4\sigma^4}{n^2}\frac{\sin^2 n\lambda}{\sin^2\lambda} && \lambda \neq 0, \pi \\
&= 8\sigma^4 + \frac{4\kappa_4}{n} && \lambda = 0, \pi \\
\text{cov}\,\{I_n(\lambda_1)I_n(\lambda_2)\} &= \frac{4\kappa_4}{n} + \frac{4\sigma^4}{n^2}\left\{\left(\frac{\sin\frac{1}{2}n(\lambda_1+\lambda_2)}{\sin\frac{1}{2}(\lambda_1+\lambda_2)}\right)^2 + \left(\frac{\sin\frac{1}{2}n(\lambda_1-\lambda_2)}{\sin\frac{1}{2}(\lambda_1-\lambda_2)}\right)^2\right\}
\end{aligned}\right\} \quad (3)$$

For λ_1 and λ_2 of the form $2\pi j/n$ the terms involving trigonometric functions vanish.

Thus two important facts emerge.

(*a*) The periodogram does not give a consistent estimation of $f(\lambda)$ since its variance is $O(1)$.

(*b*)
$$\left.\begin{aligned}\text{cov}\,\{I_n(\lambda_1)I_n(\lambda_2)\} &= O(n^{-2}) \quad \text{if } \kappa_4 = 0 \\ &= O(n^{-1}) \quad \text{if } \kappa_4 \neq 0\end{aligned}\right\}\lambda_1 \neq \lambda_2$$

It is also worth noting that the variance of $I_n(0)$ is twice that of (λ) for $\lambda \neq 0$ (neglecting terms which are $o(1)$).

We proceed now to investigate the variance of $I_n(\lambda)$ for a linear process

$$x_t = \sum_0^\infty \alpha_j \varepsilon_{t-j}$$

where ε_t satisfies the conditions on which the last derivation was based, save that we put $\sigma^2 = 1$. Put

$$J_n(\lambda, x) = \sqrt{\frac{2}{n}} \sum_1^n x_t e^{it\lambda}$$

$$= \sum_0^\infty \alpha_j e^{ij\lambda} J_n(\lambda, \varepsilon) + \sqrt{\frac{2}{n}} \sum_0^\infty \alpha_j g_{j,n}(\lambda, \varepsilon) \tag{4}$$

where

$$g_{j,n}(\lambda, \varepsilon) = \left[\sum_{\max\left\{\substack{0 \\ j-n}\right.}^{j-1} e^{i(j-k)\lambda} \varepsilon_{-k} - \sum_0^{\min\left\{\substack{n-1 \\ j-1}\right.} e^{i(n+j-k)\lambda} \varepsilon_{n-k} \right]$$

$$\left[\mathscr{E}\left\{ \left| \sum_0^\infty \alpha_j g_{j,n}(\lambda, \varepsilon) \right|^2 \right\} \right]^{1/2} \leqslant \sum_0^\infty |\alpha_j| [\mathscr{E}\{|g_{j,n}(\lambda, \varepsilon)|^2\}]^{1/2}$$

$$\leqslant \sqrt{2} \sum_0^\infty |\alpha_j| j^{1/2}.$$

Similarly $\left\{ \mathscr{E} \left| \sum_0^\infty \alpha_j g_{j,n}(\lambda, \varepsilon) \right|^4 \right\}^{1/4}$ is bounded by $K \sum_0^\infty |\alpha_j| j^{1/2}$. Thus the second and fourth absolute moments of the last term in (4) are respectively $O(n^{-1})$ and $O(n^{-2})$, uniformly in λ, if $\sum_0^\infty |\alpha_j| j^{1/2}$ converges (which will certainly be so if $\alpha_j = o(j^{-3/2})$). Under this condition we therefore have, in an obvious notation,

$$I_n(\lambda) = I_n(\lambda, x) = \left| \sum_0^\infty \alpha_j e^{ij\lambda} \right|^2 I_n(\lambda, \varepsilon) + O(n^{-1/2})$$

$$= 2\pi f(\lambda) I_n(\lambda, \varepsilon) + O(n^{-1/2})$$

where by this we mean that the neglected quantities have mean square which is of order n^{-1}. It is evident that, if the $(4k)$th moment of the ε_t is finite, then for $\sum_0^\infty | \alpha | j^{1/2} < \infty$,

$$\mathscr{E}\{| I_n(\lambda) - 2\pi f(\lambda) I_n(\lambda, \varepsilon) |^{2k}\}$$

is of order n^{-k}.

Example

The periodograms for the two examples of section II.3 are shown in figs. 3 and 4 below. In the case of the series for which the process generating the observations is known the spectral density is also shown. The periodogram was originally computed by M. G. Kendall for *periods* of 2 to 50 time intervals. Since the frequency, λ, is given by $p^{-1}2\pi$ where p is the period, the higher frequencies are sparsely represented. On the other hand when p is large, adjacent periodogram ordinates are separated by an interval (of frequencies) which is small relative to n^{-1} so that they will be relatively highly correlated. Computation of the periodogram for equally spaced periods is not in general recommended since it results in too many ordinates being computed in one range of frequencies and too few in another. Of course, in some cases one may, by reason of the physical context, be interested in periods and not frequencies and will want to compute the estimate for those periods.

In both cases shown in the diagrams the periodogram has a wildly fluctuating appearance consistent with its covariance properties.

2. Consistent estimation of the spectral distribution and density functions

We have seen that $I_n(\lambda)$ does not give a consistent estimator of $f(\lambda)$.

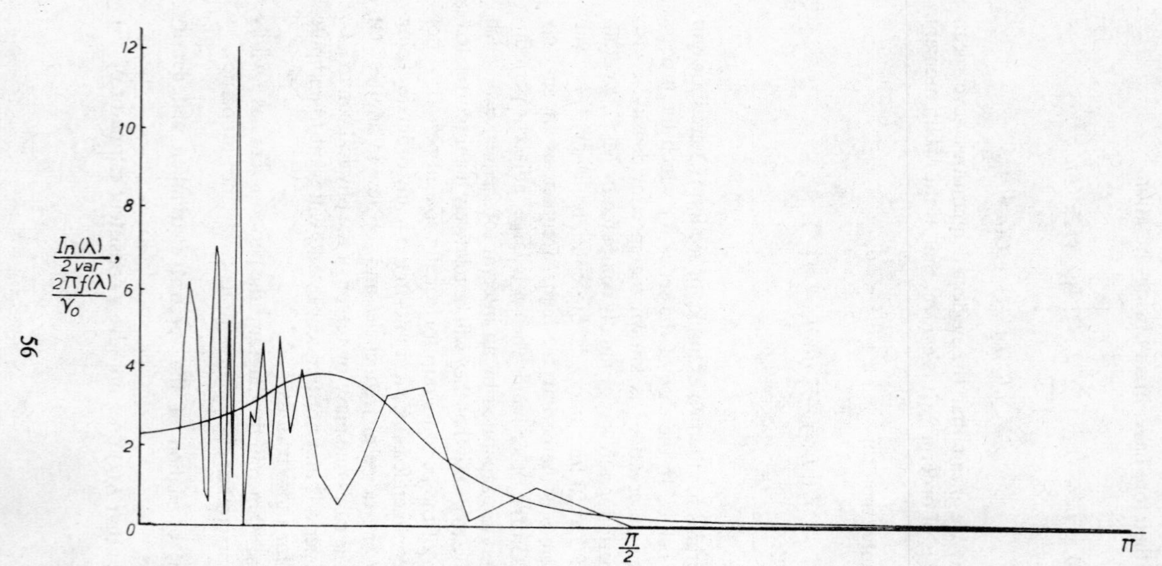

Fig. 3. Periodogram of constructed autoregressive series of example 1, section II.3. The periodogram is shown for periods from 2 to 30 and 35, 40, 45, 50 time units. The periodogram has been divided by 2 var(x_t) *so that it corresponds to* $\frac{2\pi f(\lambda)}{\gamma_0}$ *which is also shown*

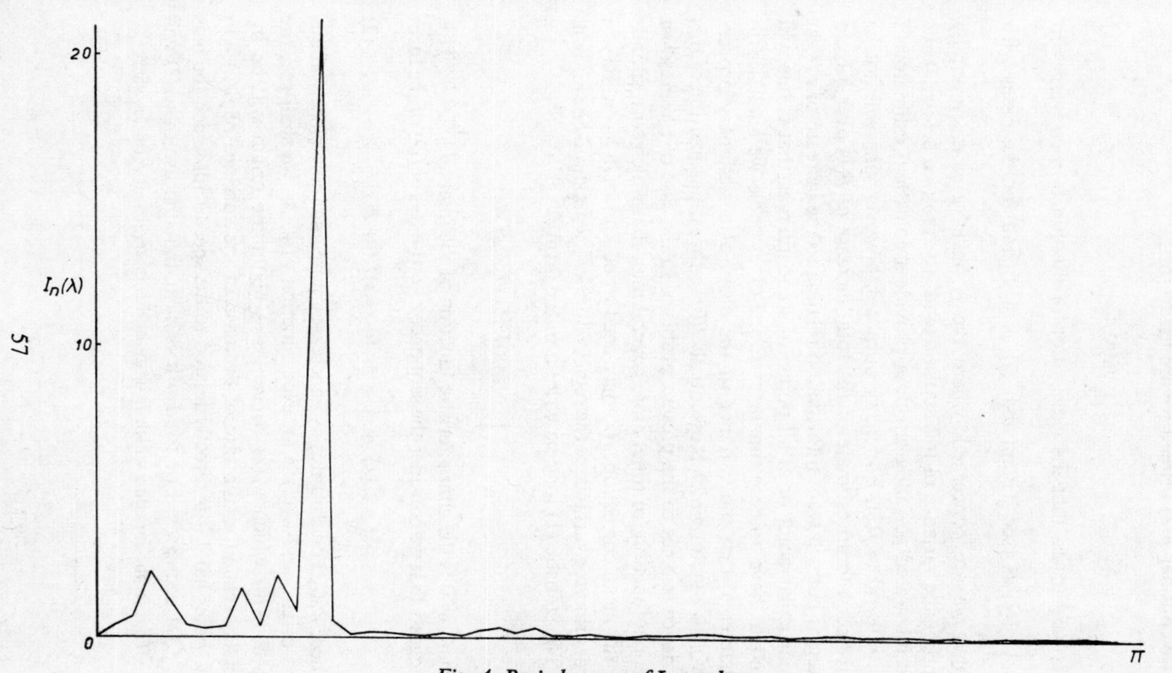

Fig. 4. Periodogram of Lynx data

If we consider $F(\lambda)$ instead we are led to form

$$\int_{\lambda_1}^{\lambda_2} I_n(\lambda)\, d\lambda. \tag{1}$$

It is easy to see that the expectation of this function converges to $4\pi \int_{\lambda_1}^{\lambda_2} f(\lambda)\, d\lambda$ (see Titchmarsh [2] p. 419) and for the cases with which we are concerned, where the c_t have variances given by (II.3.2), the variance of this statistic tends to zero (see below). Thus the integrated periodogram does provide a consistent estimator of the integral of $f(\lambda)$. For many purposes, however, one will not be satisfied with an estimate of the total increase of $F(\lambda)$ over a fixed interval but will want information relating to the increase of $F(\lambda)$ at a particular point or at all points of some interval, that is one will need to have some estimate of $f(\lambda)$. For example we shall see in Chapter V that the variance of the mean of a stationary process involves a factor which is essentially $f(0)$ so that the estimation of this variance reduces to the present problem. In the case of a regression upon a function (of time) more general than that which corresponds to the mean correction, $f(\lambda)$ may need to be known in its entirety in order that the variance of the regression coefficients should be known.

The statistic (1) may be written in the form

$$\int_{-\pi}^{\pi} w(\lambda) I_n(\lambda)\, d\lambda$$

where $w(\lambda)$ is the characteristic function of the interval $[\lambda_1, \lambda_2]$.[1] This suggests that we consider the sequence of estimates of $f(\lambda)$ of the form

$$\hat{f}_n(\lambda) = \int_{-\pi}^{\pi} w_n(\theta - \lambda) I_n(\theta)\, d\theta \tag{2}$$

where $w_n(\lambda)$ concentrates as n increases more and more closely about the origin, and $w_n(\lambda)$ is an even function of λ. As we shall see, the speed with which $w_n(\lambda)$ concentrates about the origin will be inversely related to the rate of decrease of the variance of $\hat{f}_n(\lambda)$. On the other hand it is directly related to the speed with which the bias in $\hat{f}_n(\lambda)$ decreases. For fixed n it is clear that the use of $w_n(\lambda)$ will

[1] That is, the function which is unity in this interval and zero elsewhere.

result in an estimate having a mean value smoother than the true $f(\lambda)$ so that some 'resolution' will have been lost. One needs to weigh these two effects against each other. A convenient method of doing so is to consider the mean square error

$$\mathscr{E}\{(\hat{f}_n(\lambda) - f(\lambda))^2\} = \operatorname{var}(\hat{f}_n(\lambda)) + \{B(\hat{f}_n(\lambda))\}^2$$

where

$$B(\hat{f}_n(\lambda)) = \mathscr{E}(\hat{f}_n(\lambda)) - f(\lambda).$$

Though arbitrary this combination of the two effects has an obvious appeal and we shall use it in what follows.

If

$$w_n(\lambda) \sim \frac{1}{8\pi^2} \sum_{-\infty}^{\infty} w_{n,t} e^{it\lambda}$$

then

$$\hat{f}_n(\lambda) = \frac{1}{2\pi} \sum_{-n+1}^{n-1} \left(1 - \frac{|t|}{n}\right) c_t w_{n,t} e^{it\lambda}. \tag{3}$$

For the purposes of digital computation it is this formula which will be used, not (2). It is thus preferable to begin by prescribing the $w_{n,t}$. Let us put

$$w_{n,t} = k(b_n t) \tag{4}$$

where *$k(x)$ is a bounded even function satisfying* $\int_{-\infty}^{\infty} k^2(x)\, dx < \infty$, $k(o) = 1$, *and b_n is a sequence of positive numbers converging to zero but satisfying $nb_n \longrightarrow \infty$.* We shall illustrate this by some examples. The terms by which we shall refer to these estimates are indicated in parentheses.

(i) (Truncated).

$$\begin{aligned} k(x) &= 1 \quad |x| \leqslant 1 \\ &= 0 \quad |x| > 1 \end{aligned}$$

Putting $b_n = m_n^{-1}$ we have

$$\hat{f}_n(\lambda) = \frac{1}{2\pi} \sum_{-m_n}^{m_n} \left(1 - \frac{|t|}{n}\right) c_t e^{it\lambda}$$

$$w_n(\lambda) = \frac{1}{8\pi^2} \frac{\sin \dfrac{2m_n + 1}{2}\lambda}{\sin \frac{1}{2}\lambda}.$$

This may be negative so that there is a non-zero probability that $\hat{f}_n(\lambda)$ will be negative also, which is not pleasant. The idea behind this estimate is fairly obvious. We have truncated the sequence c_t at the point $t = m_n$ leaving out the later c_t whose coefficient of variation is large. Unless the corresponding γ_t are zero this will introduce a bias of course. However, as n increases (so that $n^{-1}m_n = (nb_n)^{-1}$ tends to zero) the bias will fall. As we shall see the sequence can be chosen so that the variance decreases to zero also.

(ii) (Bartlett).[1] Bartlett introduced the estimate

$$\hat{f}_n(\lambda) = \frac{1}{2\pi}\sum_{-m_n}^{m_n}\left(1 - \frac{|t|}{m_n}\right)c_t e^{it\lambda}.$$

He obtained this by considering the break-up of the sequence of n observations into (say) p successive sequences of m_n each. If for each of these p sequences the periodogram was constructed and then the p periodograms were averaged the resulting estimate should, as both p and m_n were increased, be both asymptotically unbiased and consistent. The average of these p would give a formula such as that already written down but with c_t replaced by

$$\frac{1}{p(m_n - t)}\sum_{j=0}^{p-1}\sum_{s=0}^{m_n-t} x_{s+jm_n}x_{s+jm_n+t}$$

Since certain cross products are missing as compared with c_t it is preferable to use the latter in the final formula.

Bartlett's formula cannot be put in the form (3) with w_n, given by (4). We therefore modify it to

$$\hat{f}_n(\lambda) = \frac{1}{2\pi}\sum_{-m_n}^{m_n}\left(1 - \frac{|t|}{n}\right)\left(1 - \frac{|t|}{m_n}\right)c_t e^{it\lambda}$$

so that
$$k(x)\begin{cases} = 1 - |x| & |x| \leqslant 1 \\ = 0 & |x| > 1 \end{cases}$$

and
$$w_n(\lambda) = \frac{1}{8\pi^2}\left(\frac{\sin^2 \frac{1}{2}m_n\lambda}{m_n \sin^2 \frac{1}{2}\lambda}\right)$$

which is non-negative. That the additional term introduced into the

[1] See Bartlett[4].

modified definition will be asymptotically negligible may be seen from the derivations given below.

(iii) (Daniell).

$$k(x) = \frac{\sin \pi x}{\pi x}$$

$$f_n(\lambda) = \frac{1}{2\pi} \sum_{-n+1}^{n-1} \left(1 - \frac{|t|}{n}\right) c_t \frac{\sin(\pi b_n t)}{\pi b_n t} e^{it\lambda}$$

$$w_n(\lambda) \begin{cases} = \dfrac{1}{8\pi^2 b} & |\lambda| \leqslant \pi b_n \\ = \quad 0 & |\lambda| > \pi b_n \end{cases}$$

Thus $$f_n(\lambda) = \frac{1}{2\pi b_n} \int_{-\pi b_n}^{\pi b_n} \frac{1}{4\pi} I_n(\theta - \lambda)\, d\theta.$$

While the motivation is again clear it is evident that from the point of view of digital computations this estimator is unacceptable since all c_t must be computed, a heavy task if n is large.

(iv) (Hamming).

$$k(x) \begin{cases} = 0{\cdot}54 + 0{\cdot}46 \cos \pi x & |x| \leqslant | \\ = \quad 0 & |x| > | \end{cases}$$

$$f_n(\lambda) = \frac{1}{2\pi} \sum_{-m_n}^{m_n} \left(1 - \frac{|t|}{n}\right) c_t \left\{0{\cdot}54 + 0{\cdot}46 \cos \frac{\pi t}{m_n}\right\} e^{it\lambda}$$

$$w_n(\lambda) = \frac{1}{8\pi^2} \left[0{\cdot}54 \frac{\sin \frac{2m_n + 1}{2} \lambda}{\sin \frac{1}{2}\lambda} + 0{\cdot}23 \left\{ \frac{\sin \frac{2m_n + 1}{2}\left(\lambda + \frac{\pi}{m_n}\right)}{\sin \frac{1}{2}\left(\lambda + \frac{\pi}{m_n}\right)} + \frac{\sin \frac{2m_n + 1}{2}\left(\lambda - \frac{\pi}{m_n}\right)}{\sin \frac{1}{2}\left(\lambda - \frac{\pi}{m_n}\right)} \right\} \right]$$

(v) (Hanning).

$$k(x) \begin{cases} = \frac{1}{2}(1 + \cos \pi x) & |x| \leqslant | \\ = \quad 0 & |x| > | \end{cases}$$

This is a convenient approximation to (iv).

The reasons for these last two choices of weight function can be seen by a consideration of fig. 5, in which the weight functions (ii), (iii) and (iv) are exhibited.[1] The Bartlett weight function has a very slightly 'chunkier' main lobe, which is an advantage since most weight is then given to frequencies near to that to be estimated. The Daniell weight function is ideal from this point of view. The side

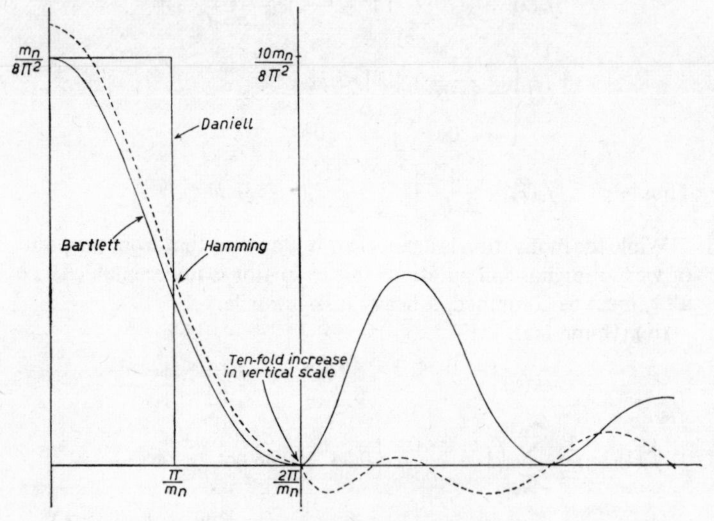

Fig. 5. Fourier Transforms of k(x)

lobes of the Bartlett weight function *eventually* decrease faster than those of the Hamming weight function (though this is not so evident in the figure). This is desirable also of course. However, the first two side lobes and particularly the first, are much higher for the Bartlett than for the Hamming weight function. The high side lobes can be a particularly unpleasant feature if $f(\lambda)$ has a sharp peak, for then subsidiary smaller (and spurious) peaks will be produced in the estimate at all frequencies for which a side lobe coincides with the

[1] Actually we have shown the Fourier transform of $k(x)$ rather than that of $k(b_n t)$. The error involved will be small if b_n is small.

main peak. The Hamming weight function is an attempt to compromise between the features discussed here. Of course the Daniell estimate does this perfectly; its weight function is as concentrated as possible with no side lobes. However, as we have seen, from a computational point of view, digitally, it is not useful.

For a discussion of these and other weight functions, from an illuminating point of view, Blackmann and Tukey [1] can be recommended.

Now let us consider the bias and variance of an estimate of the general type just considered, when

$$x_t = \sum_0^\infty \alpha_j \varepsilon_{t-j} \text{ with } \sum_0^\infty |\alpha_j| < \infty$$

and the ε_t are independent with zero mean and finite fourth cumulant. They now, of course, satisfy $\sum_0^\infty |\gamma| < \infty$. The bias involved in weighting down later c_t clearly depends upon the speed with which the γ_t converge to zero. It is therefore convenient to consider an additional restriction in the form[1]

$$\sum_{-\infty}^{\infty} |t^q \gamma_t| < \infty \quad \text{for some } q > 0. \tag{5}$$

We are not suggesting that it would be feasible to check such an exact criterion for a particular q, or that one could know *a priori* the largest q for which it will be satisfied. However, any smoothing operation on $I_n(\lambda)$ must depend upon some prior hypothesis concerning the smoothness of $f(\lambda)$ for its justification (that is on the speed with which the γ_t tend to zero) and the criterion (5) is a convenient mathematical formulation of such an hypothesis.

We shall presume that having fixed q on prior grounds, believing

[1] We, substantially, follow Parzen [1].

that (5) holds, we choose $k(x)$ so that

$$k_q = \lim_{x\to 0} \frac{1 - k(x)}{|x|^q} \tag{6}$$

exists and is finite. If this is so for some q_0 then the limit will evidently be zero for $q < q_0$.

The remaining specification required to complete the formula (3) is that of the sequence b_n. Evidently b_n^{-1}, which we think of in terms of m_n, will have to increase more slowly than n so that (as we have already mentioned) we require $nb_n \to \infty$. If $q > 1$, however, we still require, more strongly, that $nb_n^q \to \infty$, for reasons which will be apparent from the following. We shall prescribe the b_n more precisely later.

It now follows that

$$b_n^{-q}B(\hat{f}_n(\lambda)) \to -\left\{\frac{k_q}{2\pi}\sum_{-\infty}^{\infty} |t|^q\gamma_t e^{it\lambda}\right\} < \infty. \tag{7}$$

Indeed the left-hand side can be written as the sum of three terms

$$-\frac{b_n^{-q}}{2\pi}\left[\sum_{|t|\geqslant n} \gamma\, e^{it\lambda} + \sum_{-n+1}^{n-1} \gamma_t\{1 - k(b_n t)\}e^{it\lambda} + \sum_{-n+1}^{n-1} \frac{|t|}{n}\gamma_t k(b_n t)e^{it\lambda}\right]. \tag{8}$$

The middle term in (8) is

$$-\frac{1}{2\pi}\sum_{-n+1}^{n-1} |t|^q\gamma_t e^{it\lambda}\left\{\frac{1 - k(b_n t)}{|b_n t|^q}\right\} \to -\frac{k_q}{2\pi}\sum_{-\infty}^{\infty} |t|^q\gamma_t e^{it\lambda}.$$

The first and third terms in (8) converge to zero, which establishes (7). Consider, for example, the third term, when $q > 1$. Its absolute value is

$$\left| b_n^{-q}\sum_{-n+1}^{n-1} \gamma_t \frac{|t|}{n} k(b_n t)e^{it\lambda}\right| \leqslant Mb_n^{-q}n^{-1}\sum_{-n+1}^{n-1} |t\gamma_t|$$

where $|k(x)| \leqslant M$. As the summation term converges because of (5) and the fact that $q > 1$ we see that this expression converges to zero, since $nb_n^q \to \infty$.

Turning to the covariance we have

$$\lim_{n\to\infty} nb_n \operatorname{cov}\{\hat{f}_n(\lambda_1), \hat{f}_n(\lambda_2)\}$$

$$= \lim_{n\to\infty} \frac{nb_n}{4\pi^2} \sum_{\tau}\sum_{t=-n+1}^{n-1} k(b_n t)k(b_n \tau)$$

$$\times \left(1 - \frac{|t|}{n}\right)\left(1 - \frac{|\tau|}{n}\right) e^{i(t\lambda_1+\tau\lambda_2)} \operatorname{cov}(c_t, c_\tau). \quad (9)$$

We replace $n^{-1}(n - |t|)(n - |\tau|) \operatorname{cov}(c_t, c_\tau)$ by its three components (see (II.3.3)). We may then replace the term involving $K(s, t, \tau - t)$ by its limit $\kappa_4\gamma_t\gamma_\tau$ so that the contribution to (9) is

$$\lim_{n\to\infty} \kappa_4 \frac{b_n}{4\pi^2} \sum_{\tau}\sum_{t} k(b_n t)k(b_n \tau)\gamma_t\gamma_\tau e^{i(t\lambda_1+\tau\lambda_2)} = 0.$$

The second term in (II.3.3) leads to the contribution

$$\lim_{n\to\infty} b_n \sum_{\tau}\sum_{t} k(b_n t)k(b_n \tau) e^{i(t\lambda_1+\tau\lambda_2)} \beta(t + \tau). \quad (10)$$

$$\beta(s) = \frac{1}{2\pi} \int_{-\pi}^{\pi} f^2(\lambda) e^{is\lambda}\, d\lambda.$$

Rearranging (10) we obtain

$$\lim_{n\to\infty} \sum_{v=-2n+2}^{2n-2} \beta(v) e^{iv\lambda_2} \left\{ b_n \sum_{-n+1+|v|}^{n-1-|v|} k(b_n t)k(b_n(v - t)) e^{it(\lambda_1-\lambda_2)} \right\}.$$

Since the term within braces converges to

$$\lim_{m\to\infty} \int_{-\infty}^{\infty} k^2(t) e^{imt}\, dt = 0 \quad \lambda_1 \neq \lambda_2$$

$$\int_{-\infty}^{\infty} k^2(t)\, dt \qquad \lambda_1 = \lambda_2$$

and is uniformly bounded the limit (10) is

$$\begin{cases} \text{for } \lambda_1 = \lambda_2 = \lambda: & \displaystyle\int_{-\infty}^{\infty} k^2(t)\,dt \sum_{-\infty}^{\infty} \beta(v)e^{iv\lambda} \\ & \displaystyle = f^2(\lambda)\int_{-\infty}^{\infty} k^2(t)\,dt \\ \text{for } \lambda_1 \neq \lambda_2: & \text{zero.} \end{cases}$$

The remaining term in (II.3.3) gives the same non-zero contribution when $\lambda_1 = -\lambda_2$ and zero otherwise. Thus since $f(\lambda)$ is an even function we obtain

$$nb_n \operatorname{cov}\{\hat{f}_n(\lambda_1), \hat{f}_n(\lambda_2)\} \longrightarrow \begin{Bmatrix} \text{zero} & \lambda_1 \neq \pm\lambda_2 \\ f^2(\lambda)\displaystyle\int_{-\infty}^{\infty} k^2(t)\,dt & \lambda_1 = \pm\lambda_2 = \lambda \neq 0 \\ 2f^2(\lambda)\displaystyle\int_{-\infty}^{\infty} k^2(t)\,dt & \lambda_1 = \lambda_2 = 0, \pm\pi \end{Bmatrix} \quad (11)$$

To minimize the order of the mean square error we must choose b_n so that the variance and the square of the bias decrease to zero at the same rate since any change from this must slow the convergence to zero of one of them. This requires that in the limit $a_1 nb_n = a_2 b_n^{-2q}$, which implies $b_n = O(n^{-1/(1+2q)})$. The conditions $nb_n \longrightarrow \infty, q \leqslant 1; nb_n^q \longrightarrow \infty, q > 1$ are evidently then satisfied and it follows from (7) and (11) that the mean square error is, optimally, $O(n^{-2q/(1+2q)})$.

Though these results theoretically give some guide as to how b_n should be chosen, once we have fixed $k(x)$ and q so that (5) (we believe) and (6) are satisfied they appear to be of slight practical value. For example, for $q = 2$, b_n must be $O(n^{-1/5})$, which is a meaningless prescription unless n is very large since b_n^{-1} will rarely be less than 10 in any case. Nevertheless the results are important since they show that a sequence of estimates of $f(\lambda)$ may be obtained whose mean square error tends to zero at an optimal rate depending on the rate of decrease of the γ_t. Also (11) gives asymptotic formulae for the covariance matrix of the estimates.

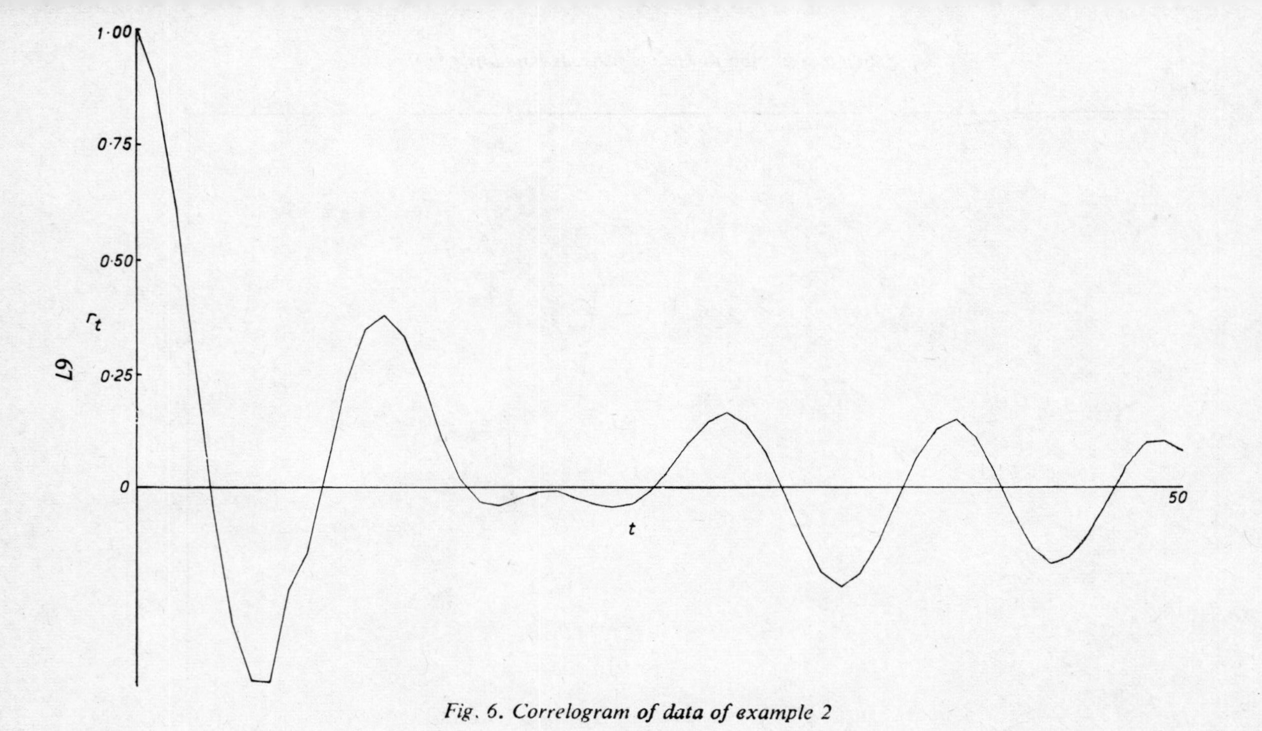

Fig. 6. Correlogram of data of example 2

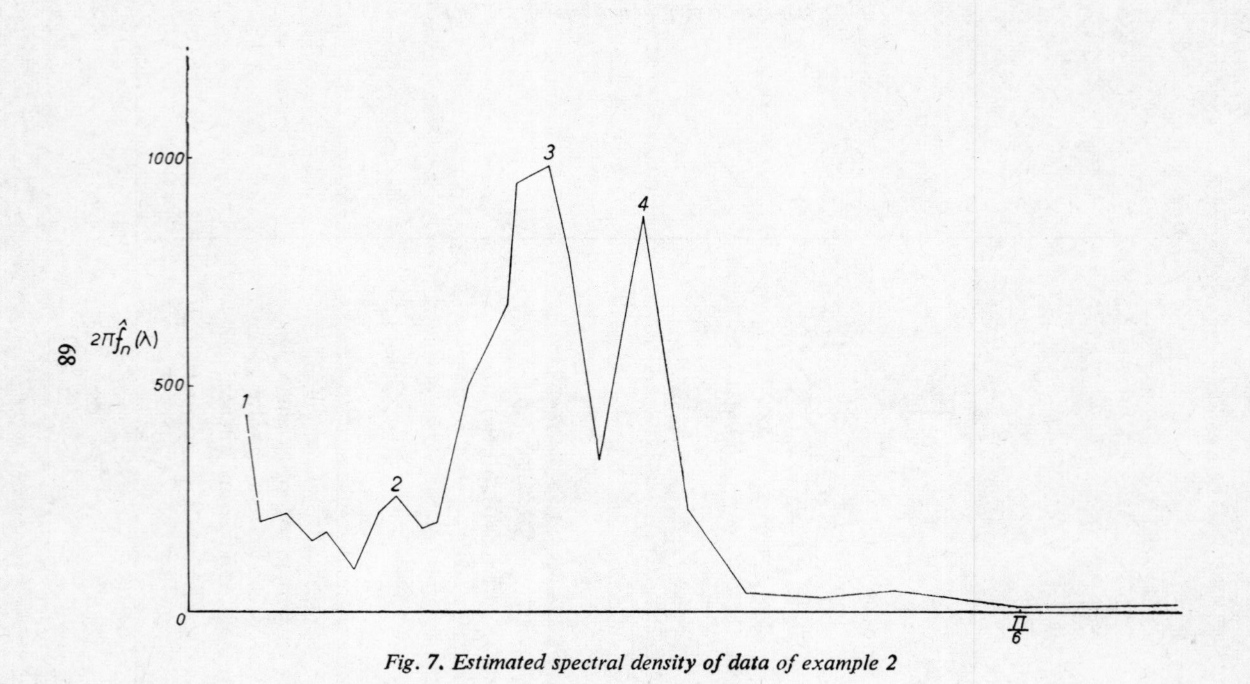

Fig. 7. Estimated spectral density of data of example 2

For later use we here consider the expressions

$$W_{i,n} = \int_{-\pi}^{\pi} w_i(\lambda) I_n(\lambda)\, d\lambda, \quad w(\lambda) \sim \frac{1}{2\pi} \sum_{-\infty}^{\infty} w_{i,t} e^{it\lambda}, \quad i = 1, \ldots, p$$

where the $w_i(\lambda)$ do not now depend on n and the x_t are as above, though (5) is dropped. If the $w_i(\lambda)$ are even functions

$$\left.\begin{aligned} \mathscr{E}\{W_{i,n}\} &\rightarrow 4\pi \int_{-\pi}^{\pi} w_i(\lambda) f(\lambda)\, d\lambda = W_i \\ n \operatorname{cov}\{W_{i,n}, W_{j,n}\} &\rightarrow (4\pi)^3 \int_{-\pi}^{\pi} w_i(\lambda) w_j(\lambda) f^2(\lambda)\, d\lambda + \kappa_4 W_i W_j \end{aligned}\right\} \quad (12)$$

For example the last term in (II.3.3) gives a contribution to the covariance which is, in the same way as above,

$$4\kappa_4 \sum_t \sum_\tau w_{i,t} w_{j,\tau} \gamma_t \gamma_\tau$$

and this converges to $\kappa_4 W_i W_j$ by the convolution theorem.

3. Some more practical considerations[1]

We have already seen that weight functions of the forms (iv) or (v) of last section have something to recommend them from the practical viewpoint of maximum concentration combined with minimal effect from secondary lobes and reasonable computations.

In any case one will compute only $m_n = b_n^{-1}$ values of $\hat{f}(\lambda)$. For weight functions (i), (ii), (iv) and (v) this is evidently all that should be done as only m_n serial covariances will be used and we shall thus be losing no information by doing just this. These may as well be evenly spaced over the interval $[0, \pi]$, say at the points $\pi j/m_n$, $j = 1, \ldots, m_n$. The shape of the weight function in fig. 5, together with the results of the first two sections of this chapter, shows that, though adjacent estimates will be fairly highly correlated, those two apart, particularly for weight functions (iii), (iv) and (v), will not. Thus if one wishes to resolve points whose distance apart, in frequencies, is d then one will need to take m_n so that (approximately)

[1] This section leans heavily upon the work of Blackmann and Tukey [1].

$\pi/m_n < \frac{1}{2}d$, i.e. $m_n > 2\pi/d$. Such an estimate will have coefficient of variation[1] whose square is, approximately,

$$\frac{\operatorname{var}(\hat{f}_n(\lambda))}{(f(\lambda))^2} \rightarrow \frac{m_n}{n}\int_{-\infty}^{\infty} k^2(x)\,dx$$

so that fractional errors in $\hat{f}_n(\lambda)$ of the order $\left\{\frac{m_n}{n}\int_{-\infty}^{\infty} k^2(x)\,dx\right\}^{\frac{1}{2}}$ are likely.

If this is too large one will have either, (1) to obtain more observations or, (2) sacrifice some resolution or, (3) give up the project.

A number of other, more or less practical, considerations are discussed below.

(*a*) It is evident that for a process of independent random variables the effect of smoothing will vanish and there will be no bias. If one knew *a priori* the nature of the spectral density one could of course reduce oneself, at least approximately, to this most satisfactory situation by choosing a trigonometric polynomial

$$\sum_{-q}^{q} \delta_j e^{ij\lambda} = \left|\sum_{0}^{q} \alpha_j e^{ij\lambda}\right|^2$$

which is uniformly near to $f(\lambda)^{-1}$, and by then replacing the original series x_t by the series

$$y_t = \sum_{0}^{q} \alpha_j x_{t-j} \quad t = q+1, \ldots, n.$$

The new series y_t will have a spectral density

$$\left|\sum_{0}^{q} \alpha_j e^{ij\lambda}\right|^2 f(\lambda)$$

which will be nearly uniform and so may be most satisfactorily estimated. One may then re-convert the estimate to an estimate of $f(\lambda)$ by division by $\left|\sum_{0}^{q} \alpha_j e^{ij\lambda}\right|^2$. This procedure may be looked at

[1] We presume that $f(\lambda)$ is not zero in the interval $[0, \pi]$.

from the view of an attempt to choose a weight function, $w(\lambda)$, to suit the special needs of the $f(\lambda)$ being estimated.[1]

Of course one does not know $f(\lambda)$ *a priori*. However, one could, having carried out an initial estimate of this function, use this estimate to suggest an approximation to $f(\lambda)^{-1}$ in the form of a trigonometric polynomial and proceed as before. If one had the time and energy and the project was sufficiently important the operation could be repeated again. The initial estimate could, for example, be based on a fairly low value of m to save computations.

(*b*) If the true process has a non-zero mean then this must be eliminated since otherwise a 'spurious' jump in the spectral function will be introduced at the origin of frequencies. We shall deal with this problem in Chapter V but mention it here since raw data will almost always need to have, at least, the effect of the mean removed before treatment.

(*c*) Finally we may consider the situation which arises when a continuous time record is replaced by a discrete series of observations at equidistant points $j\Delta$; $j = 0, 1, \ldots$ We have, for the continuous time series,

$$\gamma(t)e^{it\lambda} = \int_{-\infty}^{\infty} e^{it\theta} f(\theta - \lambda)\, d\theta.$$

Then from Poisson's summation formula,[2] we have

$$\sum_{-\infty}^{\infty} \left\{ f\left(\lambda - j\frac{2\pi}{\Delta}\right) \right\} = \frac{\Delta}{2\pi} \sum_{-\infty}^{\infty} \gamma(j\Delta) e^{ij\Delta\lambda}.$$

The serial covariances, c_j, which we shall compute will be estimates of $\gamma(j\Delta)$ (though we shall rescale time so that $\Delta = 1$) so that the spectral density we shall be estimating will be

$$f_\Delta(\lambda) = \sum_{-\infty}^{\infty} f\left(\lambda - j\frac{2\pi}{\Delta}\right), \quad \frac{-\pi}{\Delta} \leqslant \lambda \leqslant \frac{\pi}{\Delta}.$$

[1] The expression (11) above shows that this procedure does not affect the variance of the estimate asymptotically. Since it will clearly reduce the bias, it is a worthwhile procedure.

[2] See Titchmarsh [1] p. 61. We presume that $f(\lambda)$ is sufficiently regular for the relation to hold.

The estimate we shall get will, therefore, be more or less vaguely related to the function which takes on the value $f(\lambda)$ in the interval $[-\pi/\Delta, \pi/\Delta]$, depending upon the degree to which frequencies outside of this range are important. The range can of course be made arbitrarily long by a choice of Δ sufficiently small.

We conclude this section with two examples.

Example 1

We consider the data of example 1 of section 3, Chapter II. Here 480 observations are available, a comparatively large number. The spectral density has a peak at about $\pi/5$. If a tendency for the data to oscillate with that frequency was noticed[1] one would probably want to separate this point from the origin (which in practical analysis is always likely to be affected by 'spurious' effects due to trends). This leads us to require

$$m \geqslant \frac{2\pi}{\pi/5} = 10.$$

For safety we take $m = 15$. The coefficient of variation will now be, approximately

$$\left\{\frac{\int_{-\infty}^{\infty} k^2(x)\,dx}{32}\right\}^{1/2}.$$

For the weight functions we shall use we have

Truncated: $\int_{-\infty}^{\infty} k^2(x)\,dx = 2$

Bartlett: ,, $= 2/3$

Hanning: ,, $= 3/4$.

Thus to this order of approximation the coefficient of variation will not exceed $\frac{1}{4}$ for any of these three. We shall not evaluate the biases. Evidently the truncated estimate will have the smallest bias ($k_q = 0$ for all finite q for this weight function) and in fact it will be very small in the present instance, for this weight function, since for

[1] This is not very likely as the peak is by no means sharp, though an examination of the correlogram of the second 60 observations in the series (see Kendall [1]) shows that it is by no means impossible.

$j > 15$ the true serial correlations are quite negligible, while the factor $(1-|j|/480)$ will be very near to unity. The three estimates are compared with the true values in Table 2 below. In this table we have shown $\gamma_0^{-1}2\pi f(\pi j/15)$ for $j = 0, \ldots, 14$ and the corresponding estimates. In the estimates a further bias will have been introduced by the division by the estimate, c_0, of γ_0, but this will not affect the comparison of the general shape of the curves.

TABLE 2

j	$\frac{2\pi}{\gamma_0} f\left(\frac{\pi j}{15}\right)$	Truncated	Bartlett	Hanning
0	2·263	2·170	2·802	3·024
1	2·486	3·873	3·085	3·151
2	3·209	2·690	2·804	2·935
3	3·646	2·677	2·507	2·699
4	2·458	2·395	2·040	2·043
5	1·168	0·706	1·095	1·132
6	0·573	0·714	0·686	0·597
7	0·318	0·225	0·407	0·346
8	0·198	0·150	0·290	0·163
9	0·136	0·126	0·176	0·114
10	0·100	0·082	0·153	0·104
11	0·079	0·131	0·130	0·082
12	0·067	−0·001	0·093	0·058
13	0·059	−0·009	0·090	0·050
14	0·055	−0·018	0·077	0·037

None of the estimates shows the peak near $\pi/5$. A tendency to a peak near $\pi/15$ is shown instead. Even for the truncated estimate the differences do not appear to be significant. The largest, for this, is just over two standard deviations (fractional error of $\frac{1}{2}$) and this is of course the greatest deviation in 15. The effect is not due to the smoothing formulae tending to cut across (above) the trough at the origin of frequencies and (below) the peak at $\pi/5$ since, as has been observed, the bias is small. In fact the result is to be expected after an examination of the correlogram of section 3 of Chapter II and the periodogram of section 1 of this chapter. This latter shows some very large values near $\pi/20$ but only smaller ones around $\pi/5$.

The tendency to a peak near $\pi/15$ is, partly at least, due to a false trough at the origin caused by the removal of the mean.[1]

The effect of the coincidence of side (inner) lobes with peaks is not very evident though it can be seen in the truncated estimate at $j = 6$ and 11 (the side lobes are at a peak approximately $\pi/6$ and $\pi/3$ away from the peak of the weight function $w(\lambda)$).

Of course in this example one would do much better if one guessed that the process was a second order autoregression, as one can see by considering the first example in section II.4. In general for processes whose spectral densities are of the present type (having only one relative maximum) one will almost certainly do better by fitting some such finite parameter scheme as an autoregression. For this reason it may be simpler to plot the correlogram first. If this appears to have a very simple structure one may fit an autoregression to begin with and subsequently examine the spectral density of the residuals[2] thus avoiding a considerable amount of computation.

Of course if one followed the recommendation (*a*) of this section one would probably be led by the shape of any of the three estimated curves to try a second order autoregression, which would be that estimated in example 1 of section II.4, and the analysis of the residuals by any of the methods discussed would certainly lead to a very nearly uniform spectrum and thus a final estimate of $f(\lambda)$ very near to the correct one.

In cases where the spectral density is very complicated, however, one will hardly be able to guess its nature at all accurately from the correlogram, as the following example shows.

Example 2

This example is taken from Whittle [3] and relates to 660 observations of the water level in a channel on the coast of New Zealand. The 'Bartlett' estimate was used with $m = 114$. This will completely resolve points differing by about $(114)^{-1}2\pi$. The coefficient of variation will be about 0·34. The correlogram is shown in fig. 6 and the estimated spectral density in fig. 7. In this case also, apparently for reasons connected with the physical context, the estimate $\hat{f}_n(\lambda)$ was

[1] See Chapter V below.

[2] See also the following chapter.

computed for integral periods. However, the corresponding frequencies are separated by an interval which makes their correlation low and though the higher frequencies are sparsely represented this hardly matters since this part of the spectrum is in this case very unimportant. There seems no doubt of the reality of the peaks numbered 1, 3 and 4 in fig. 7. The immediate neighbourhood of the origin has been omitted since a trend in the data[1] had been removed by means of a second degree polynomial. The spectrum is certainly fairly complicated and could only be approximated adequately by a very high order autoregression. It is more difficult, in this case, to discern the detail in the correlogram. The individual serial correlations do not correspond to 'independent' (orthogonal) aspects of the data so that their interpretation, except in the case of simple structures, is not so immediate.

[1] See Chapter V.

CHAPTER IV

Hypothesis Testing and Confidence Intervals

1. Testing for a jump in the spectral function

In the last two chapters we have considered problems of estimating the structure of a stationary process, either on the assumption that the underlying stochastic scheme was some finite parameter scheme or on some, perhaps, less well formulated assumption, concerning the smoothness of $f(\lambda)$. In both cases we put aside the possibility of a jump in $F(\lambda)$. Clearly this cannot be put aside, however, and we must consider the problem of determining whether there are jumps in this function as well as the problem of removing their effects so that the structure of the part of $\{x_t\}$ corresponding to the a.c. component of $F(\lambda)$ may be estimated by the foregoing methods. We shall consider the first part of this problem in this section and the second part in the last chapter.

First let us consider the problem of testing for jumps in $F(\lambda)$, *on the null hypothesis that the observations are independent and Gaussian.* If the point λ_0, at which the jump may occur, is fixed by the alternative hypothesis the problem is a simple one solved by standard methods of regression and correlation analysis. It is of course just the problem of testing the multiple correlation of x_t with $\cos t\lambda_0$ and $\sin t\lambda_0$. We consider therefore the further problem of testing whether jumps are present in $F(\lambda)$, without being able, *a priori*, to specify where these happen.

We consider the periodogram ordinates $I_n(\psi_j)$, $j = 0, 1, \ldots, [\frac{1}{2}n]$, where the square bracket indicates that we take the integral part of the indicated number ($\frac{1}{2}n$ if n is even, $\frac{1}{2}(n - 1)$ if n is odd). We see immediately from (III.1.2) and (III.1.3) that the $\gamma_0^{-1} I_n(\psi_j)$ are independently distributed as chi-square with two degrees of freedom (i.e. $\gamma_0^{-1}I_n(\psi_j)$ is $\chi^2_{(2)}$) with the exception of $\gamma_0^{-1}I_n(\psi_0)$ and, for n even, $\gamma_0^{-1}I_n(\psi_{\frac{1}{2}n})$ whose distributions have only one degree of freedom. It is simpler to omit the last two from our considerations. For $I_n(\psi_0)$ this is no loss for in practice we shall not know that x_t has a zero mean and shall have to use $x_t - \bar{x}$. This will not affect $I_n(\psi_j)$ for $j \neq 0$ but makes $I_n(\psi_0)$ zero. The omission of $I_n(\psi_{\frac{1}{2}n})$ (for n even) will not affect the test we are about to give (in power) except for situations where the jump in $f(\lambda)$ occurs at or very near to π.

We shall put $m = [\frac{1}{2}(n-1)]$ and replace the $I_n(\psi_j)$, $j = 1, \ldots, m$, by the ratios

$$g_j = I_n(\psi_j) \Big/ \sum_{j=1}^{m} I_n(\psi_j) = y_j \Big/ \sum y_j$$

(let us say) which will be distributed independently of γ_0. A suitable test statistic against the hypothesis of a jump in $F(\lambda)$ will be the greatest of the g_j. However, we may wish to test for further jumps once one or more is established, so that we are led to consider the distribution of the rth greatest of the g_j.[1] We shall renumber the g_j so that

$$g_1 \geqslant g_2 \geqslant \ldots \geqslant g_r \geqslant \ldots g_m.$$

Remembering that the $\gamma_0^{-1}y_j$ are independent $\chi^2_{(2)}$ and that, since y_j is distributed independently of γ_0, we may put $\gamma_0 = 2$, we derive the characteristic function of g_r^{-1} as

$$\mathscr{E}\{e^{i\theta/g_r}\} = m\binom{m-1}{r-1}\int_{y_r=0}^{\infty}\int_{y_{r-1}=y_r}^{\infty}\cdots\int_{y_1=y_r}^{\infty}\int_{y_{r+1}=0}^{y_r}\cdots\int_{y_m=0}^{y_r} e^{i\theta\frac{\Sigma y_j}{y_r}-\Sigma y_j}\,\Pi dy,$$

where the first factor arises (*a*) from the fact that any one out of the m may be the rth greatest, and (*b*) the ways of choosing the $(r-1)$ out of the remaining $(m-1)$ which are to be greater than this rth

[1] The distribution of the greatest g_j was found by Fisher [1]. The derivation for the rth greatest given here is due to Whittle [1].

greatest one. This expression is easily evaluated as

$$\frac{m!}{(m-r)!(r-1)!}\int_0^{\infty}\frac{e^{r(i\theta-y_r)}\{1-e^{i\theta-y_r}\}^{m-r}}{\left(1-\frac{i\theta}{y_r}\right)^{m-1}}dy_r.$$

It is easily seen that this is an (absolutely) integrable function of θ for $m > 2$ so that we may invert the Fourier transform, to obtain the frequency function of g_r^{-1}, as

$$(g_r^{-1})$$

$$=\frac{1}{2\pi}\frac{m!}{(m-r)!(r-1)!}\int_{-\infty}^{\infty}e^{-i\theta/g_r}\int_0^{\infty}\frac{e^{r(i\theta-y_r)}\{1-e^{(i\theta-y_r)}\}^{m-r}}{(1-i\theta/y_r)^{m-1}}dy_r d\theta$$

$$=\frac{r}{2\pi}\binom{m}{r}\sum_0^{m-r}(-)^j\binom{m-r}{j}\int_{-\infty}^{\infty}\int_0^{\infty}\frac{e^{-i\theta/g_r}e^{(i\theta-y_r)(j+r)}}{(1-i\theta/y_r)^{m-1}}dy_r\,d\theta. \quad (1)$$

Since the integrand is absolutely integrable we may reverse the order of integration and first evaluate

$$\int_{-\infty}^{\infty}\frac{e^{-i\theta/g_r}e^{i\theta(j+r)}}{(1-i\theta/y_r)^{m-1}}d\theta.$$

If $j + r > 1/g_r > 1$ this integral vanishes (since the pole at $\theta = -iy_r$ lies in the bottom half plane). If $j + r < 1/g_r$ the last expression is

$$-2\pi y_r^{m-1}e^{y_r(j+r-g_r^{-1})}(g_r^{-1}-j-r)^{m-2}/(m-2)!$$

Inserting this in (1) and evaluating the integral with respect to y_r we obtain the frequency function of g_r (adjoining a factor g_r^{-2}) as

$$\frac{m!(m-1)}{(m-r)!(r-1)!}\sum_{j=r}^{[1/g_r]}(-)^{j-r}\binom{m-r}{j-r}(1-jg_r)^{m-2}.$$

We finally obtain the probability that $g_r > x$ (by integrating from x to $1/r$, which is the maximum possible value of g_r) as

$$P\{g_r > x\} = \frac{m!}{(r-1)!}\sum_r^{[1/x]}(-)^{j-r}\frac{1}{j(m-j)!(j-r)!}(1-jx)^{m-1}. \quad (2)$$

For $r = 1$ and 2 a table of significance points is given in Fisher [2] and a more extensive table, for $r = 1$, in Fisher [1]. Over the range there considered ($m \leqslant 50$) an adequate approximation is obtained by taking the first term only of (2), that is by using the approximation

$$P\{g_r > x\} \approx \binom{m}{r}(1 - rx)^{m-1}.$$

The test as it stands has a restricted scope. The most outstanding difficulty lies, perhaps, in the fact that the null hypothesis is that of randomness. It is quite likely that one will wish to test the hypothesis of a jump in $F(\lambda)$ against the null hypothesis of an absolutely continuous, but not uniform, $F(\lambda)$. If the null hypothesis is specified completely *a priori* and $f(\lambda)$ has no zeros in $[-\pi, \pi]$ we are led by the consideration of section III.1 to form the quantities

$$K_n(\psi_j) = I_n(\psi_j)/\{2\pi f(\psi_j)\}$$

and to proceed, using these, as we would use the $I_n(\psi_j)$ if the conditions of the first part of this discussion were satisfied. *We still presume that the ε_t are normal.* Then, under the conditions of section III.1, it is not hard to show[1] that

$$P\left\{\max_j \mid K_n(\psi_j) - I_n(\psi_j, \varepsilon) \mid > \delta_n\right\} < \delta_n$$

where $\lim_n \delta_n = 0$.

We introduce the notations K_r and $g_r(\varepsilon)$, respectively, for the rth greatest of the quantities $K_n(\psi_j) \Big/ \sum_j K_n(\psi_j)$ and

$$I_n(\psi_j, \varepsilon) \Big/ \sum I_n(\psi_j, \varepsilon).$$

[1] This may be proved, for example, by using a multivariate form of Tchebycheff's inequality (Olkin and Pratt [1]), applied to $\{K_n(\psi_j) - I_n(\psi_j, \varepsilon)\}^2$ which, as we have seen in section III.1, has mean less than $k_1 n^{-1}$ and variance less than $k_2 n^{-2}$.

If $d_n(\alpha)$ satisfies

$$P\{g_r(\varepsilon) > d_n(\alpha)\} = \alpha$$

we see from the relation

$$\frac{K_n(\psi_j)}{d_n(\alpha)\sum K_n(\psi_j)} = \left\{\frac{I_n(\psi_j, \varepsilon)}{d_n(\alpha)\sum I_n(\psi_j, \varepsilon)} + \frac{K_n(\psi_j) - I_n(\psi_j, \varepsilon)}{d_n(\alpha)\sum I_n(\psi_j, \varepsilon)}\right\}\left\{1 + \frac{\sum_j (K_n(\psi_j) - I_n(\psi_j, \varepsilon))}{\sum_j I_n(\psi_j, \varepsilon)}\right\}^{-1}$$

that $K_r/d_n(\alpha)$ converges in probability to $g_r(\varepsilon)/d_n(\alpha)$. This follows from the fact that $m^{-1}\sum I_n(\psi_j, \varepsilon)$ will converge in probability to σ^2 and $md_n(\alpha)$ will increase indefinitely while

$$\max | K_n(\psi_j) - I_n(\psi_j, \varepsilon) |$$

converges in probability to zero.

Thus

$$P\{K_r > d_n(\alpha)\} \longrightarrow \alpha.$$

Thus the use of the $K_n(\psi_j)$ is justified, at least asymptotically. However, even in this form the test is of little use as $f(\lambda)$ will rarely be completely prescribed. Whittle [4] has suggested that we should replace $f(\psi_j)$ by some estimate; obtained either from a smoothing procedure by the methods of Chapter III or by estimating the unknown parameters in some prescribed finite parameter form of $f(\lambda)$ by the methods of Chapter II.[1]

We shall consider only the case of an autoregressive null hypothesis (normal ε_t). Then we replace, in the $K_n(\psi_j)$,

$$\left|\sum_0^p \beta_k e^{ik\psi_j}\right|^2 \quad \text{by} \quad \left|\sum_0^p \hat{\beta}_k e^{ik\psi_j}\right|^2$$

[1] A possible alternative procedure to Whittle's has been briefly indicated by M. B. Priestley in the discussion (p. 435) at the most recent [3] of the Royal Statistical Society's Symposia on Time Series.

to obtain a set of statistics which we shall call $\hat{K}_n(\psi_j)$. However

$$\mathscr{E}\left\{\max_j \left| \sum_l \sum_k (\beta_k\beta_l - \hat{\beta}_k\hat{\beta}_l)e^{i(k-l)\psi_j} \right|\right\}$$
$$\leqslant \left\{\sum_l \sum_k \mathscr{E} \mid \beta_k\beta_l - \hat{\beta}_k\hat{\beta}_l \mid\right\} \rightarrow 0$$

(since the $\hat{\beta}_k$ converge in mean square to the β_k) so that the maximum error involved in replacing $2\pi f(\psi_j)$ by its estimate converges in probability to zero. A repetition of the previous type of argument establishes the asymptotic validity of the test based on the maximum $\hat{K}_n(\psi_j)$. In the case of a smoothed estimate of $f(\psi_j)$ the discussion would need to be more delicate since one cannot reduce it to a consideration of a fixed finite number of parameters.

There are some problems associated with the test which remain.

(*a*) The results which have been derived have been based upon the requirement of normality for the ε_t. Asymptotically it appears that this will not matter, for the test is effectively based upon the quantities $mI_n(\psi_j, \varepsilon) \Big/ \sum_j I_n(\psi_j, \varepsilon)$ which will converge in probability to $\sigma^{-2}I_n(\psi_j, \varepsilon)$. The greatest of these should have a distribution which, under fairly general conditions, will be independent of the distribution of the $I_n(\psi_j, \varepsilon)$ (see for example Kendall [2] vol. 1 p. 217). However, this asymptotic situation may be approached only very slowly. One expects that, even for n not very large, the distribution of the criterion K_r will not be much affected by departures from normality since $I_n(\psi_j, \varepsilon) \Big/ \sum_j I_n(\psi_j, \varepsilon)$ has its first two moments independent of the distribution of the ε_t, to the first order. This difficulty, that nothing much is known about the speed of approach of the asymptotic situation, is one which, at the present moment, is encountered throughout the theory of time series.

(*b*) The most serious difficulty is associated with the power of the test. This may be reduced by two factors. The first is the possibility that a jump in $F(\lambda)$ will occur at a point between the points ψ_j. If the jump occurs at a point λ midway between two of these points then the contribution of the corresponding harmonic to the nearest $I_n(\psi_j)$ will be about 2/5ths of its contribution to $I_n(\lambda)$ (see Whittle [1] p. 107). This difficulty is of an essential nature and cannot be entirely avoided without prior information. A more important difficulty arises from the use of an estimate of $f(\lambda)$. This is very evident in the case of a smoothed estimate since if a jump occurs, at a point λ_p, this estimate will have a marked peak in the neighbourhood of the point of the jump. A similar difficulty will be met with in the case of a fitted autoregression, which will also be certain to give an estimated $f(\lambda)$ with a marked peak near λ_p (at least if the autoregression is of the second order or higher). Of course if there are a number of points of jump and a fairly low order autoregression is chosen, on prior grounds, peaks will not occur at all of the points of jump. Our theory has of course presumed that the *form* of $f(\lambda)$ is chosen on prior grounds. However, this is hardly likely to be so in practice. Again the difficulty is an essential one since on the basis of a reasonable number of observations one can never hope to discriminate well between the possibilities of a jump in $F(\lambda)$ and a sharp peak in $f(\lambda)$.

Example 1

Let us reconsider the Lynx data of section III.3. We consider the null hypothesis of a second order autoregression and the alternative of a second order autoregression plus an harmonic component. The largest value of $\{2\pi\hat{f}(\psi_j)\}^{-1}I_n(\psi_j)$ (which is $k \times 0{\cdot}558$) is at $j = 33$, corresponding to quite a minor peak in $I_n(\psi_j)$ (see section III.1), the effect of the marked peak in $\hat{f}(\psi_j)$ having been to very greatly reduce the peak in $I_n(\psi_j)$ at $j = 12$. The sum of the $\{2\pi\hat{f}(\psi_j)\}^{-1}I_n(\psi_j)$ is $k \times 7{\cdot}322$. Thus the ratio is 0·076 (the factor k being $\hat{\sigma}^2$). Thus

$$P\{g_1 > 0{\cdot}076\} \approx 57(1 - 0{\cdot}076)^{56} = 0{\cdot}670.$$

Thus the largest ratio is not significant at any reasonable level. The peak at $j = 33$ is not isolated, the ordinate at $j = 34$ being nearly as large. In these circumstances a reasonable procedure is to test the second largest peak on the grounds that it is the two together which will constitute significance. Here

$$g_2 = 0{\cdot}509/7{\cdot}322 = 0{\cdot}0695$$

$$P\{g_2 > 0{\cdot}0695\} = 57 \times 28(1 - 0{\cdot}319)^{56} = 0{\cdot}37$$

which is still far from significant.

Example 2

We consider next the second example of section III.3. Whittle fitted an autoregression to this data (by methods to be discussed below) and found the relation

$$x_t - 1{\cdot}50x_{t-1} + 0{\cdot}47x_{t-2} + 0{\cdot}40x_{t-3} - 0{\cdot}18x_{t-4} = \hat{\varepsilon}_t.$$

He investigated lags of up to nine time periods but after the first four the remainder added nothing significant to the explanation. The largest ratio $\{2\pi\hat{f}(\psi_j)\}^{-1}I_n(\psi_j)$ occurred at $j = 5$ [1] and its ratio to the sum of these is 0·0418. Now

$$P\{g_1 > 0{\cdot}0418\} \approx 330(1 - 0{\cdot}0418)^{330} = 0{\cdot}00025$$

which is highly significant.

A second peak, near $j = 61$, was also isolated and found significant (see Whittle [3]).

2. Tests relating to autoregressive schemes

Our results of section II.4 already provide us with a large sample theory for testing the significance of the coefficients in an hypothesized autoregression when the residuals are independent random variables with finite fourth moment. Indeed these results show that we may proceed in the case of an autoregression exactly as we should for the classic case of a regression of one variable on a set of variables entirely distinct from it, when the residuals from that regression are normally and independently distributed, and our test procedures will be, at least asymptotically, valid.

Example

Consider again the data of example one of section II.3. Let us test the hypothesis that the data are generated by an autoregression of degree 2 against the hypothesis that the degree of the autoregression

[1] The precise frequency was not quite a multiple of $(660)^{-1}2\pi$, the true relative maximum in $I_n(\lambda)$ having been located. The error involved by this procedure will be too small to affect the validity of the test.

is one. In this case 480 observations are available so that the asymptotic theory should be sufficiently precise. The estimate of β_2 is $\hat{\beta}_2 = 0{\cdot}485$ and we know from section II.4 that $\hat{\beta}_2$ is approximately normal with variance given by the bottom right-hand element of $(n - p)^{-1}\sigma^2\mathbf{\Gamma}_p^{-1}$ which is estimated by the appropriate element of $(n - p)^{-1}\hat{\sigma}^2\mathbf{G}_p^{-1}$ as $0{\cdot}0016$. Since $\hat{\beta}_2$ is about twelve times its estimated standard deviation there is no doubt of its significance. This test is, of course, asymptotically equivalent to the formation of the partial serial correlation between x_t and x_{t-2} with the effects of x_{t-1} eliminated. If we indicate this coefficient by the symbol $r_{02\cdot1}$ we see easily that

$$r_{02\cdot1} \approx (r_2 - r_1^2)/(1 - r_1^2)$$

and may be tested as an ordinary correlation with n degrees of freedom, if n is large.

(*a*) *A simple Markoff alternative*

We proceed in this section to obtain some more precise results on the basis of stricter conditions. We shall henceforth assume that the x_t are generated by a Gaussian process (with zero mean). We consider first the test of the hypothesis that the x_t are independent against the hypothesis that they are generated by a first order autoregression (simple Markoff process). We shall rename β_1 as $-\rho$ in this case to accord with the usual convention. The near sufficiency of r_1 as an estimator of ρ leaves no doubt that the test statistic should be close to this statistic.[1] It should also be pointed out that it is usual to use a one-sided test of such an hypothesis (so that only large positive values of r_1 are considered as significant) since a negative ρ is rarely met in practice. A very large number of papers[2] have been written upon the distribution of r_1, or modifications of it, the initial work being that of Vòn Neumann [1]. Von Neumann obtained the distribution of

$$\frac{\delta^2}{s^2} = \frac{1}{n-1}\sum_{1}^{n-1}(x_{t+1} - x_t)^2 \Bigg/ \frac{1}{n}\sum_{1}^{n}(x_t - \bar{x})^2.$$

[1] For a full discussion see T. W. Anderson [1].

[2] For references see Watson [1], Daniels [1] and Jenkins [2].

It is easily seen that

$$\frac{\delta^2}{s^2} = \frac{2n}{n-1}\left\{1 - \frac{\frac{1}{2}(x_1-\bar{x})^2 + \frac{1}{2}(x_n-\bar{x})^2 + \sum_{2}^{u}(x_t-\bar{x})(x_{t-1}-\bar{x})}{\sum_{1}^{n}(x_t-\bar{x})^2}\right\}$$

$$= \frac{2n}{n-1}\{1 - r_A\}$$

let us say. From this the distribution of r_A may be found and a comparison of the 1% points of this distribution, for a one-sided test (see T. W. Anderson [1]), with those for an ordinary correlation (with mean corrections) from $n + 3$ observations shows that for $n \geqslant 20$ the statistic r_A may be treated as an ordinary correlation from $n + 3$ observations. For smaller n the exact distribution of $s^{-2}\delta^2$, tabulated by Hart [1],[1] could be consulted (though the 1% point, even for $n = 10$, is only slightly changed, the exact figure being 0·624 instead of 0·634).

Another statistic whose distribution has been obtained under the present conditions (see R. L. Anderson [1]) is

$$r_1' = \sum_{1}^{n}(x_t - \bar{x})(x_{t-1} - \bar{x}) \Bigg/ \sum_{1}^{n}(x_t - \bar{x})^2$$

$$= \frac{\mathbf{x}'\mathbf{W}_1\mathbf{x} - n\bar{x}^2}{\mathbf{x}'\mathbf{x} - nx^2}.$$

This is called the first circular serial correlation coefficient. The proper values of the numerator quadratic forms are; for δ^2/s^2, $2(1 - \cos\psi_j)$; for r_1', $\cos\psi_j$; $j = 0, \ldots, n-1$. In the case of δ^2/s^2 the proper values are all distinct whereas, for n odd, the proper values of r_1', save for the zero proper value, occur in pairs. This makes the distribution of the latter statistic easier to obtain since the joint characteristic function of the numerator and denominator has only poles and no branch points (see below). For $n = 2m + 1$ we therefore derive the distribution of r_1' (following Whittle [1]). If

[1] See also T. W. Anderson [1].

the components of the vector $\mathbf{x}$ using the proper vectors of $\mathbf{W}_1$ as basis are $z_0, z_1, \ldots, z_{n-1}$, then we easily see that

$$r_1' = \sum_1^{n-1} \cos \psi_j z_j^2 \Big/ \sum_1^{n-1} z_j^2$$

$$= \sum_1^{n-1} \mu_j z_j^2 \Big/ \sum_1^{n-1} z_j^2 = \frac{\xi}{\zeta}$$

where the z_j are again, of course, $N(0, 1)$ [1] and independent. The joint characteristic function of the numerator and denominator is seen to be

$$\phi(\theta_1, \theta_2)$$

$$= \frac{1}{(2\pi)^{(n-1)/2}} \int_{-\infty}^{\infty} \ldots \int \exp - \tfrac{1}{2}\left\{\sum_1^{n-1} z_j^2(1 - 2i\mu_j\theta_1 - 2i\theta_2)\right\} \prod dz_j$$

$$= \prod_1^m (1 - 2i\mu_j\theta_1 - 2i\theta_2)^{-1}.$$

Inverting, we form first (for $m > 2$)

$$\frac{1}{2\pi} \int_{-\infty}^{\infty} \exp \{-i(\theta_1\xi + \theta_2\zeta)\} \prod_1^m (1 - 2i\mu_j\theta_1 - 2i\theta_2)^{-1} \, d\theta_1.$$

Taking residues at the relevant poles (those corresponding to μ_j for which $\mu_j\xi > 0$ since for $\mu_j\xi \leqslant 0$ the poles may be avoided by integrating around a contour in the opposite half plane[2]) we obtain the expression

$$\sum_p{}' A_p \exp \left\{-\tfrac{1}{2} \frac{\xi}{\mu_p}(1 - 2i\theta_2)\right\}(1 - 2i\theta_2)^{-(m-1)} e^{-i\theta_2\zeta}$$

where $$A_p = \frac{1}{2\mu_p} \prod_{\substack{j \neq p \\ j=1}}^{m} \left(1 - \frac{\mu_j}{\mu_p}\right)^{-1}$$

[1] That is, normal with zero mean and unit variance.

[2] For $\xi = 0$ we take, of course, all of those residues corresponding to $\mu_j < 0$.

and the summation runs over those p for which $\mu_p\xi > 0$. Then the joint density of ξ and ζ is

$$f(\xi, \zeta) = \sum_p{}' A_p e^{-\frac{1}{2}\zeta} \frac{1}{2\pi} \int_{-\infty}^{\infty} e^{\frac{1}{2}(\zeta-\xi/\mu_p)(1-2i\theta_2)}(1 - 2i\theta_2)^{-(m-1)} \, d\theta_2$$

(see Whittaker and Watson [1] p. 246) where the summation is now over those p for which $\mu_p\xi > 0$ *and* $(\zeta - \xi/\mu_p) > 0$ (since for $\zeta - \xi/\mu_p \leqslant 0$ the pole at $\theta_2 = -\frac{1}{2}i$ may be avoided by integrating around a contour in the opposite half plane). Thus the distribution of r_1' is

$$\frac{1}{2^{m-1}\Gamma(m-1)} \sum{}'' A_p\left(1 - \frac{r_1'}{\mu_p}\right)^{m-2} \int_0^{\infty} e^{-\frac{1}{2}\zeta} \zeta^{m-1} \, d\zeta$$
$$= (n-3) \sum{}'' A_p\left(1 - \frac{r_1'}{\mu_p}\right)^{\frac{1}{2}(n-5)}$$

where the summation is now over those p for which $r_1'\mu_p > 0$ [1] and $\zeta - \xi/\mu_p > 0$, i.e. $r_1'/\mu_p < 1$. For r_1' positive this is

$$\frac{n-3}{2} \sum_1^s (\cos\psi_p - r_1')^{\frac{1}{2}(n-5)} \prod_{j=1}^{\frac{1}{2}n-1}{}' (\cos\psi_p - \cos\psi_j) \qquad (1)$$

for $\cos\psi_{s+1} \leqslant r_1' \leqslant \cos\psi_s$ and the prime indicates that the zero factor is omitted. Some significance points of this distribution are tabulated in R. L. Anderson [1]. It may be mentioned that a comparison of the 1% point of the two distributions shows that for $n > 20$ one may take $r_1' + n^{-1}$ as being distributed as an ordinary correlation from $n + 3$ observations (and again with reasonable accuracy for $n \geqslant 10$). The term n^{-1} is nearly the difference in the expectations of r_A and r_1'. (On the null hypothesis the mean of r_1' is n^{-1} to this order.) It may be preferable to use r_1 as defined by (II.3.4) and treat it in the same way (i.e. use $r_1 + n^{-1}$ as an ordinary correlation from $n + 3$ observations) since the expectation of r_1 increases with ρ slightly faster than that of r_1'. The two statistics will in practice differ only very slightly.[2]

[1] $\mu_p < 0$ if $r_1' = 0$.

[2] As a further example of the distribution theory of statistics related to r_1 we mention that the ordinary correlation between the series x_{2t} and $x_{2t-1} + x_{2t+1}$,

(b) Autoregression of higher order

Once the null hypothesis ceases to be that of independence an essential difficulty is encountered which makes the labour involved in determining the distributions very considerable. For now the frequency function of the observations on the null hypothesis becomes

$$(2\pi)^{-n/2} \mid \boldsymbol{\Gamma}_n \mid^{-\frac{1}{2}} \exp \{-\tfrac{1}{2}\mathbf{x}'\boldsymbol{\Gamma}_n^{-1}\mathbf{x}\}$$

In order to derive the distributions of the partial (or partial multiple) serial correlations the joint distribution of all of the r_j (up to a certain order) will be needed. Thus one will be involved in finding the joint distribution of a set of ratios of quadratic forms $(\mathbf{x}'\mathbf{x})^{-1}\mathbf{x}'\mathbf{Q}_j\mathbf{x}$. The derivation will certainly be simplified if the $\mathbf{Q}_j$ can be simultaneously reduced to diagonal form, which requires that they should commute.[1] This could always be achieved for example by choosing the $\mathbf{Q}_j$ to be the circulants

$$\mathbf{W}_j = \tfrac{1}{2}(\mathbf{U}^j + \mathbf{U}'^j)$$

where $\mathbf{U}$ is defined by (I.2.1). We shall call the ratios for this choice of $\mathbf{W}_j$ the circular serial correlation coefficients and indicate them by r_j'. However, the derivation will still be difficult unless $\boldsymbol{\Gamma}_n$ commutes with the $\mathbf{W}_j$. If x_t is generated by an autoregression one will not be able to find a suitable set of $\mathbf{Q}_j$ which commute with $\boldsymbol{\Gamma}_n$. This has led to the development of the distribution theory of the r_j' on the *fictitious* assumption that the x_t are generated by a process defined on the finite group of integers reduced modulo n so that $x_{t+kn} = x_t$ for all k (integral). This has been called a circular process. We have seen in the first chapter that such processes may be chosen so as to have a spectral function (necessarily a jump function of course) uniformly close to the spectral function of (say) an autoregression. One hopes therefore that this departure from reality will result in a distribution theory which, for reasonable n, will be a fair approximation to the

$t = 1, \ldots, [\frac{1}{2}(n-1)]$ (with mean corrections) is *precisely* distributed as an ordinary correlation from $[\frac{1}{2}(n-1)]$ observations. Asymptotically the test based on this statistic is fully efficient but this may not be so for small samples. See Ogawara [1], Hannan [1].

[1] See Halmos [2] p. 141.

truth.[1] The matrix $\mathbf{\Gamma}_n$ and the $\mathbf{W}_j$ will now commute since they are all circulants. In practice what has been done is to consider a circular process generated by a relation of the form[2]

$$\sum_{0}^{p} \beta_j x_{t-j} = \varepsilon_t, \quad \beta_0 = 1$$

where the ε_t come from a circular process of independent variables. It is then easy to see that the first p r_j' together with c_0 are jointly sufficient for the β_j and σ^2 so that the joint distribution of c_0 and the $r_j (j = 1, \ldots, p)$ may be found on the hypothesis $\sigma^2 = 1$,

$$\beta_1 = \beta_2 = \ldots = \beta_p = 0$$

and the density function in the general case obtained by the adjunction of a known factor. For if $\mathbf{t}$ (a vector) is sufficient for $\boldsymbol{\theta}$ (a vector) then

$$f(\mathbf{x} \mid \boldsymbol{\theta}) = g(\mathbf{t} \mid \boldsymbol{\theta}) f_1(\mathbf{x})$$

so that $\quad g(\mathbf{t} \mid \boldsymbol{\theta}) = g(\mathbf{t} \mid \boldsymbol{\theta}_0) f(\mathbf{x} \mid \boldsymbol{\theta}) / f(\mathbf{x} \mid \boldsymbol{\theta}_0) = g(\mathbf{t} \mid \boldsymbol{\theta}_0) h(\mathbf{t}, \boldsymbol{\theta}).$

The derivation of the general distribution of the r_j' $(j = 1, \ldots, p)$ was given by Quenouille [1]. It may be obtained (see Watson [1]) by (essentially) an extension of the methods used above for r_1'. Finally it may be pointed out that if one is concerned with testing the hypothesis $\beta_p = 0$ then one shall use the partial (circular) serial correlation $r'_{0p \cdot 1, 2, \ldots, p-1}$ obtained by correlating $x_1, \ldots, x_n$ with $x_{p+1}, \ldots, x_{p+n}$, $x_{p+n} = x_p$, with the effects of the intermediate vectors removed. This statistic will, upon the hypothesis $\beta_p = 0$, be distributed independently of $\beta_1, \ldots, \beta_{p-1}$ so that from the point of view of this test these may be put equal to zero.

These distributions, while interesting, are however too complicated for practical use. The reason for the complicated nature of the distributions is seen to lie in the fact that the region of joint variation of the statistics $r_1', \ldots, r_p'$ in p space is polyhedral (having vertices

[1] We must emphasize that for the null hypothesis of independence there is no departure from reality involved.

[2] Thus we make the β_j agree rather than making the two spectral functions agree as closely as possible.

determined by the proper values of the $\mathbf{W}_j$) so that the analytic form of the distribution function changes as a vertex is passed.[1] This has led to the derivation of approximations to the distribution obtained by replacing the finite set of proper values occurring in the characteristic function by a continuous set or, equivalently, by replacing the polyhedral region by a region having a boundary free of vertices. We can illustrate the method by considering the characteristic function of ξ and ζ discovered above. This is

$$\prod (1 - 2i\mu_j\theta_1 - 2i\theta_2)^{-1}$$

$$= \exp -\frac{n}{2\pi}\left\{\frac{2\pi}{n}\sum_1^n \log(1 - 2i\mu_j\theta_1 - 2i\theta_2)\right\}$$

$$\approx \exp -\frac{n}{2\pi}\int_0^{2\pi} \log(1 - 2i\theta_2 - 2i\theta_1\cos\lambda)\,d\lambda$$

(cf. section I.5).

This integral may be evaluated and the resulting (approximate) characteristic function inverted to obtain the density function

$$\frac{\Gamma(\frac{1}{2}n+1)}{\Gamma(\frac{1}{2})\Gamma(\frac{1}{2}n+\frac{1}{2})}(1 - r_1'^2)^{\frac{1}{2}(n-1)} \quad (2)$$

(see Dixon [1]). This is the density function of an ordinary correlation (with mean corrections) from $n + 3$ observations.[2]

For $\rho \neq 0$ the distribution may also be obtained by this method and involves the adjunction of a factor to (2) (see Leipnik [1]). A slightly more accurate approximation due to Jenkins [3] and Daniels [1] is[3]

$$f(r_1' \mid \rho) = \frac{\Gamma(\frac{1}{2}n+\frac{1}{2})(1 - r_1'^2)^{\frac{1}{2}n-1}}{\Gamma(\frac{1}{2})\Gamma(\frac{1}{2}n)(1-\rho)(1 - 2\rho r_1' + \rho^2)^{\frac{1}{2}n-1}}\left\{1 - \frac{n+1}{n-1}r_1'\right\}. \quad (3)$$

[1] The same problem is encountered in the distribution derived in section 4.1 and is well brought out in Fisher's original derivation. See Fisher [1].

[2] The previously recommended procedure that $r_1' + 1/n$ (or better $r_1 + 1/n$) should be treated as an ordinary correlation with $n + 1$ degrees of freedom still remains as a slight improvement.

[3] It should not, incidentally, be thought that the difficulties with the case $\rho \neq 0$ can be avoided by considering $x_t - \rho x_{t-1} = \varepsilon_t$. The ε_t will, of course,

Moreover, Jenkins shows that the first two moments of this distribution are, to the order n^{-1}, the same as those *for the distribution of* r_1 (but not of r_1') so that this may be used as a fair approximation to the density function of this statistic. As it stands the function is not in a very easily usable form. The transformations (Jenkins [1])

$$r_1 = \sin y, \quad \rho = \sin \lambda, \quad x = y - \lambda$$

result *for the case where no mean correction is made* in x having a distribution approximately normal with mean $-\{(1-\rho^2)n^2\}^{-\frac{1}{2}}\frac{3}{2}\rho$ and variance n^{-1}. The use of these formulae for the case where a mean correction *is* made may suffice if n is not too small.

The use of the smoothing procedure does not give satisfactory results for the partial serial correlations (which become important when $j > 1$). For example if the joint density of r_1' and r_2' (on the hypothesis of randomness) is considered, and the marginal distribution of r_2' determined, this marginal distribution is not appropriate (Watson [2]). Daniels [1] and Jenkins [2], [3] obtained better approximations to the distributions of the partial (circularly defined) serial correlations on the basis of a modified smoothing procedure. If we use $r_{0p\cdot 12,\ldots p-1}$ to indicate the partial serial correlation between x_t and x_{t+p} (with mean corrections and without a circular definition) then $r_{0p\cdot 12,\ldots,p-1}$ may be satisfactorily defined by

$$r_{0p\cdot 12,\ldots p-1} = \frac{\left|\mathbf{R}_{p+1}^{1,\,p+1}\right|}{|\mathbf{R}_p|}$$

where $\mathbf{R}_p = c_0^{-1}\mathbf{G}_p$ (see section II.4) and $\left|\mathbf{R}_{p+1}^{1,\,p+1}\right|$ is the cofactor of the element in the first row and $(p+1)$st column. Thus

$$r_{02\cdot 1} = (1 - r_1^2)^{-1}(r_2 - r_1^2),$$

in agreement with our earlier definition.

---

give an exact test of a prescribed ρ and a confidence interval for ρ but the use of the data in this way is not efficient. For small ρ, however, it might be a worthwhile procedure as the asymptotic efficiency of the resulting test is $(1-\rho^2)$ which is greater than 0·91 for $\rho < 0{\cdot}30$.

Then Daniels and Jenkins show that *with a circular definition:*
(*a*) for p even

$$f(r'_{0p\cdot 1,\ldots,p-1}) \approx k(1-r'_{0p\cdot 1\ldots p-1})^2(1-r'^2_{0p\cdot 1,\ldots,p-1})^{\frac{1}{2}(n-3)} \qquad (4)$$

$$k^{-1} = B(\tfrac{1}{2}, \tfrac{1}{2}n - \tfrac{1}{2}) + B(\tfrac{3}{2}, \tfrac{1}{2}n - \tfrac{1}{2})$$

(*b*) for p odd

$$(r'_{0p\cdot 1,\ldots,p-1})$$

$$\approx B(\tfrac{1}{2}, \tfrac{1}{2}n - \tfrac{1}{2})(1 - r'_{0p\cdot 1,\ldots,p-1})(1 - r'^2_{0p\cdot 1,\ldots,p-1})^{\frac{1}{2}(n-3)} \qquad (4)'$$

where $$B(p, q) = \Gamma(p)\Gamma(q)/\Gamma(p+q).$$

By analogy with the case for r_1 and r'_1 (see the paragraph below formula (3)) it is suggested by Jenkins that these will be better approximations to the densities for $r_{0p\cdot 1,\ldots,p-1}$ than for $r'_{0p\cdot 1,\ldots,p-1}$ and thus should be used as appropriate to the former statistic.

The significance points of these distributions may be obtained from tables of the incomplete Beta function. For a test against partial serial correlation there is usually no reason, *a priori*, for expecting

TABLE 3

5% *points for two-sided test with equal tail areas*

		No. of observations		
		15	20	25
Distribution (4)	Upper Lower	0·36 −0·62	0·32 −0·54	0·31 −0·46
$r_{2\cdot} + 2/n$ as correlation from $n + 2$ observations	Upper Lower	0·35 −0·62	0·32 −0·52	0·30 −0·46

only positive values and a two-sided test will be used. The means of the statistics having the densities (4) and (4)′ are $-2/(n+1)$ and $-1/n$ respectively. The variances are $(n-1)(n+3)\{(n+1)^2(n+2)\}^{-1}$ and $n^{-2}(n-1)$. This suggests that $r_{02\cdot 1} + 2/n$ and $r_{03\cdot 12} + 1/n$ be treated as correlations from $n + 2$ observations. The distributions (4) and (4)′ are not symmetrical about their means but Table 3 above nevertheless suggests that these approximations will be sufficient for most practical purposes. In practice data will not be normal (nor precisely stationary) so that a too pedantic approach to this testing problem could hardly be justified.[1]

Example

The following observations are the first 20 of the series discussed in example 1 of section II.3.

0, 15, −26, −28, −11, 24, 41, 76, 89, 34, 40, −15, −1, 52, 30, −40, −42, −32, −59, −45.

The first three serial correlations and the partial serial correlations together with the population values are shown below.

	r_j	ρ_j	$r_{0j\cdot 1, 2, \ldots, j-1}$	$\rho_{0j\cdot 1, 2, \ldots, j-1}$
= 1	0·637	0·733		
2	0·275	0·307	−0·220	−0·500
3	0·159	−0·029	0·152	0

Here $r_1 + 1/n = 0{\cdot}687$ and tested as an ordinary correlation from 23 observations is highly significant. The first partial serial correlation can be seen from Table 3 above to be not significant and the same will evidently be true of the second partial correlation. With 20 observations only one can hardly hope to estimate the nature of the process generating the observations very precisely, of course.

3. Tests of goodness of fit

The tests which we have considered so far are tests against very particular alternatives, i.e. tests against a jump in the spectrum or against a certain autoregressive alternative. The second type of test can of course be fairly general as one could, for example, test the hypothesis of an autoregression of order h against that of order $h + k$ or less by using the corresponding partial multiple (serial) correlation.[2] However, here we wish to discuss, somewhat briefly, other tests which have been proposed as tests of 'goodness of fit' when the alternatives held in view are only fairly vaguely specified.

[1] This must not, of course, be interpreted as a denigration of the value of Daniels' and Jenkins' work as, apart from its value when n is really small, the approximations suggested above are in any case based upon their results.

[2] See also Whittle [4].

Let us return to a formula of the type

$$\Phi_{j,n} = \int_{-\pi}^{\pi} I_n(\lambda)\phi_j(\lambda)\,d\lambda$$

where $\phi_j(\lambda)$ does not now play the part of a delta function, however. The considerations of section III.2 show that the means and covariances of such statistics approach, respectively,

$$\Phi_j = 4\pi \int_{-\pi}^{\pi} f(\lambda)\phi_j(\lambda)\,d\lambda$$

and
$$n^{-1}\left[(4\pi)^3\int_{-\pi}^{\pi} f^2(\lambda)\phi_j(\lambda)\phi_k(\lambda)\,d\lambda + \kappa_4\Phi_j\Phi_k\right]. \tag{1}$$

Here we presume that x_t is generated by a linear process whose coefficients converge to zero sufficiently fast and κ_4 is the fourth cumulant of the ε_t.

If we choose $\phi_j(\lambda)$ so that Φ_j is zero and so that the $(2\pi)^{3/2}\phi_j(\lambda)$ are orthonormal with respect to $f^2(\lambda)$ then the statistics $\Phi_{j,n}$ will (*asymptotically*) have zero mean and will be uncorrelated while their variance will be n^{-1}, independently of the nature of the distribution of the ε_t. This suggests the use of

$$n\sum_{1}^{m}\{\Phi_{j,n}\}^2 \tag{2}$$

as a chi-square variate with m degrees of freedom to test the hypothesis that the true spectral density is $f(\lambda)$, if the asymptotic normality of the $\Phi_{j,n}$ can be established.

Some possible choices for the $\phi_j(\lambda)$ are shown below. The $\phi_j(\lambda)$ are there given as complex valued functions for simplicity of presentation but in each case the imaginary part is an odd function and so does not contribute to $\Phi_{j,n}$ or Φ_j. To obtain the correct expression for the variance, from (1), $\phi_k(\lambda)$ must be conjugated and the factor $\frac{1}{2}$ adjoined (or the real part of $\phi_j(\lambda)$, only, taken).

We presume that $f(\lambda) \neq 0$, a.e., in the cases where $f(\lambda)^{-1}$ occurs.

(1) $\phi_j^{(1)}(\lambda) = \frac{1}{2}\left(\frac{1}{2\pi}\right)^2\{f(\lambda)\}^{-1}e^{ij\lambda};\ j > 0$

[1] Here, and in future considerations, m is regarded as fixed as n increases.

(2) $\phi_j^{(2)}(\lambda) = \frac{1}{2}\left(\frac{1}{2\pi}\right)^2 \{f(\lambda)\}^{-1} \frac{P(e^{i\lambda})}{P(e^{-i\lambda})} e^{-ij\lambda}; \quad j > p$

where $P(z)$ is a polynomial in z with real coefficients, of degree p, having no zeros on or in the unit circle. If $f(\lambda)$ corresponds to an autoregression of order p (expressible as a moving average, in terms of past and present values of ε_t) then $P(e^{i\lambda}) = \sum_0^p \beta_j e^{ij\lambda}$ will satisfy these conditions. The first p of the c_j will not occur in the $\phi_j(\lambda)$ as just defined and may be used by orthonormalizing $(e^{-i\lambda} - \rho_j)$ with respect to $f^2(\lambda)$ by the Gram–Schmidt process.

(3) For a moving average of order q we may obtain the $\phi_j^{(3)}(\lambda)$

(*a*) for $q < j$ by orthonormalizing $e^{-ij\lambda}$ with respect to $f^2(\lambda)$ by the Gram–Schmidt process;

(*b*) for $0 < j \leqslant q$ by orthonormalizing $e^{-ij\lambda} - \rho_j$ with respect to $f^2(\lambda)$ by the Gram–Schmidt process.

For (1) and (2) if $f(\lambda)$ corresponds to an autoregression, and for (3) if $f(\lambda)$ corresponds to a moving average, the $\Phi_{j,n}$ are finite linear combinations of the c_j and thus will be asymptotically normal. However, (1) (due to Bartlett and Diananda [1]) is less useful unless the autoregression coefficients, β_j, are specified since the effects of estimation are not (asymptotically) confined to a subset of the $\Phi_{j,n}$ (which will be dropped) and are thus not easily removed. (Of course γ_0 may always be replaced by an estimate, as γ_0^{-1} will occur merely as a factor.) If the first p of the c_j and c_0 are used to estimate γ_0 and the moving average coefficients for (3) the asymptotic properties of the $\Phi_{j,n}$ $p < j$, will still hold. This case is due to Wold [2]. However, here we shall confine our attention to (2) (Quenouille [2]) for the case of an autoregression of order p, where $P(e^{i\lambda})$ has been chosen accordingly. The $\sqrt{n}\,\Phi_{j,n}$ are then easily seen to be

$$\sqrt{n}\left\{\sum_0^p\sum \beta_k\beta_l\gamma_{k-l}\right\}^{-1} \sum_0^p\sum \beta_k\beta_l c_{j-k-l}\left(1 - \frac{|j-k-l|}{n}\right), \quad p < j \tag{3}$$

whose asymptotic properties are also easily checked directly. As originally suggested by Quenouille (and in the form usually used) the factors $(1 - |j - k - l|/n)$ were not present and the factor $\sqrt{n}$ in (3) was replaced by $(n - j + p)^{\frac{1}{2}}$. As this will give a statistic which, for small samples, will have a variance nearer to unity than (3) these modifications should be made.[1] Moreover

$$n^{\frac{1}{2}} \sum_{\substack{k, l \\ =0}}^{p} \sum (\beta_k \beta_l - \hat{\beta}_k \hat{\beta}_l) c_{j-k-l} \tag{4}$$

$$= n^{\frac{1}{2}} \sum_k \left\{ (\beta_k - \hat{\beta}_k) \sum_l (\beta_l + \hat{\beta}_l) c_{j-k-l} \right\}.$$

But $\sum (\beta_l + \hat{\beta}_l) c_{j-k-l}$ is easily seen to converge in probability to zero for $j > p$ since $\sum_0^p \beta_j \gamma_{t-j} = 0$, $t > 0$. Since $n^{\frac{1}{2}}(\beta_k - \bar{\beta}_k)$ converges in probability to a normal variable with zero mean and finite variance the expression (4) converges in probability to zero.

Thus the $\hat{\Phi}_{j,n}^{(2)}$ for $j > p$ (where the circumflex has the usual meaning) may be used as in (2) (with n replaced by $n - j + p$ as suggested above) and the asymptotic theory will still hold. For $j \leqslant p$ of course the $\hat{\Phi}_{j,n}^{(2)}$ are identically zero, these degrees of freedom having been lost in the process of estimation. It is not hard to see that, for $j > p$, $\hat{\Phi}_{j,n}^{(2)}$ is equivalent to $nr^2_{0j \cdot 12 \ldots p}$ so that the chi-square statistic based on the first m of these is

$$n \sum_{p+1}^{p+m} r^2_{0j \cdot 12 \ldots p}. \tag{5}$$

This contrasts with the test statistic based on the partial multiple

[1] The original derivations in the references given are in some ways preferable to that given here as they show that if the β_j and σ^2 are known and, for example, for $\phi_j^{(1)}(\lambda)$, up to $m + p$ observations are discarded in the definitions of the c_j then the distribution theory is exact, for any n.

serial correlation of x_t with $x_{t+p+1}, x_{t+p+2}, \ldots, x_{t+p+m}$ with the effects of intermediate variates removed, which will be of the form

$$\prod_{j=p+1}^{p+m} (1 - r^2_{0j\cdot 12\ldots j-1}). \tag{6}$$

Asymptotically (5) and (6) will be equivalent (under reasonable conditions) since the products of the $r^2_{0j\cdot}$ in (6) will give terms converging in probability to zero. For small samples (5) may be preferable from the point of view of the validity of the application of the *asymptotic theory* because of the absence of these product terms. Moreover the fact that the partial correlations involved are of lower order (in the sense that only p variates are 'partialled' out compared with up to $p + m - 1$ for the partial multiple correlation) suggests that the approach to the *asymptotic* distribution theory may be quicker, though there is no very substantial evidence for this.[1] Finally the test has a decided advantage computationally over the multiple partial correlation because of the lower order of the partial correlations involved. Often one will want to test the fit of a low order autoregression ($p = 1$ or 2 say) by comparing it with autoregressions up to quite a high order (say $m = 25$). Quenouille's test, once the r_j are computed (27 in our case) can be carried through by a single computer on a hand machine in under an hour. The multiple partial correlation involves the evaluation of high order determinants (up to 27 in our case), a much more arduous procedure. As a result of these advantages of robustness, power (with respect to autoregressive alternatives) and computational convenience Quenouille's test has been widely used.

We shall briefly mention here two other tests, due to Bartlett [3], based upon the periodogram directly. As we have seen in section IV.1 the maximum difference of the statistics $f(\psi_j)^{-1}I_n(\psi_j)$ from the quantities $I_n(\psi_j, \varepsilon)$, the periodogram ordinates of the disturbances in a linear process whose coefficients decrease to zero reasonably fast and for which ε_t has finite fourth cumulant κ_4, is a quantity of order

[1] Some sampling experiments were carried out by Quenouille [2] but n in all of these was large (240 or 480). See also Bartlett and Rajalakshman [1].

$n^{-\frac{1}{2}}$. Moreover the same was shown to be true, for the autoregression at least, if $f(\lambda)$ is replaced by an estimate $\hat{f}(\lambda)$. If the ε_t are normal then $I_n(\psi_j, \varepsilon)$ is proportional to a chi-square with two degrees of freedom so that a suitable test for the null hypothesis of a specified $f(\lambda)$ (or an $f(\lambda)$ of a specified autoregressive form) will be Bartlett's [5] modification of the Neyman–Pearson test for the homogeneity of a set of variances. We consider the case where the interval $[0, \pi]$ is divided into certain fixed sub-intervals i_k, $k = 1, \ldots, s$, and we form the averages

$$s_k^2 = \frac{1}{2n_k} \sum_{i_k} I_n(\psi_j)\hat{f}(\psi_j)^{-1} \tag{7}$$

(or the same formula using $f(\lambda)$ if this is specified) where j ranges over those n_k values for which ψ_j falls within the kth sub-interval. Then s_k^2 is approximately a variance (from normal observations) based on $2n_k$ degrees of freedom. Such a grouping of the $I_n(\psi_j)\hat{f}(\psi_j)^{-1}$ is appropriate for two reasons.

(*a*) If the number of groups is allowed to increase with n it is not evident that the asymptotic theory is valid.

(*b*) The power[1] of the test for a given level of significance will depend upon the number of degrees of freedom involved and the difference between the expectation of the chi-square statistic (defined below) on the null and alternative hypothesis. This expectation will not be greatly affected by grouping frequencies over a range within which, on the *alternative hypothesis*, $\mathscr{E}\{I_n(\psi_j)f(\psi_j)^{-1}\}$ is fairly constant while this grouping will reduce the degrees of freedom involved and thus increase the power.

The test statistic will, then, be

$$\left.\begin{aligned} M &= K\left[n \log_e \left\{\frac{\sum_k 2n_k s_k^2}{n}\right\} - \sum_k 2n_k \log_e s_k^2\right] \\ K &= \left[1 + \frac{1}{3(s-1)}\left\{\sum_k \left(\frac{1}{2n_k}\right) - \frac{1}{n}\right\}\right]^{-1} \end{aligned}\right\} \tag{8}$$

which will asymptotically be distributed as chi-square with $(s - 1)$

[1] See Hannan [4].

degrees of freedom. There is no doubt that, in the case where $\hat{f}(\lambda)$ is used, this asymptotic theory will apply in a much wider range of circumstances than the case of the autoregressive null hypothesis which we have covered. The test based on the $\Phi_{j,n}^{(2)}$ (or the equivalent multiple partial serial correlation test) will clearly be preferable when an autoregressive null *and* alternative hypothesis is being considered (see the reference in the last footnote) but for the more common case where only a vague set of alternatives can be specified, and particularly for circumstances where the alternatives can best be specified in terms of possible variations in the shape of $f(\lambda)$ over certain ranges, this test has much to recommend it. Two further comments may be made.

(*a*) The computations required, if n is large, are considerable. However, if one was worried about a *particular* range of frequencies only, the test could be confined to that range and the computations reduced.

(*b*) It is not hard to check that the variance of s_k^2 converges to $\{\sigma^4 + n_k\kappa_4/n\}/n_k$ in the general case where ε_t is not necessarily normal. Thus the distribution of M cannot (asymptotically) be that of a chi-square based on $s-1$ degrees of freedom but instead depends upon the unknown κ_4 (since n_k/n will not converge to zero for groups based upon an interval of frequencies of fixed length). If all, or nearly all, of the n_k/n are small, of course, the effect will be small also. There seems little doubt that one could justify the application of the asymptotic theory for a sequence of statistics for which $n_k \longrightarrow \infty$ while $n_k/n \longrightarrow 0$.

For the second test due to Bartlett we consider the case where x_t is generated by a linear process for which $\alpha_j = o(j^{-3/2})$ and the ε_t are normal. Finally we require that $f(\lambda)$ should not be zero in $[-\pi, \pi]$. This last is needed to justify the replacement of $I_n(\psi_j, \varepsilon)$ by $\{2\pi f(\psi_j)\}^{-1} I_n(\psi_j, x)$ which will be made below.

The quantities $I_n(\psi_j, \varepsilon)$ have the density ke^{-kx} and a simple argument shows that the joint distribution of the $(m-1)$ quantities

$$u_{p/m} = \sum_1^p I_n(\psi_j, \varepsilon) \Bigg/ \sum_1^m I_n(\psi_j, \varepsilon), p = 1, \ldots, m-1, m = [\tfrac{1}{2}(n-1)]$$

has the density $(m-1)!$ where now

$$0 < u_{p/m} < u_{p+1/m} < 1.$$

Thus the quantity $u_{p/m}$, $p = 1, \ldots, m$ is the sum of the first p out of m random variables which are rectangularly distributed and independent save for the condition that their sum is unity. The event

$$\max_p \sqrt{m} \left| u_{p/m} - \frac{p}{m} \right| < \alpha$$

may be diagrammatically represented as the falling of all the heavily dotted points in fig. 8 within the region, $\mathscr{R}$, bounded by two lines, of unit slope, at a distance $\alpha/\sqrt{m}$ from the line of unit slope through the origin. Save for occurrences such as that marked with an arrow in fig. 8 this event is the same as that which requires that the whole irregular line should fall within $\mathscr{R}$. This last is, however, just the event that the empirical distribution function of m rectangularly distributed variables should differ from the theoretical distribution by no more than $\alpha/\sqrt{m}$, uniformly over the interval [0, 1]. This last probability tends as m increases (see for example Feller [1]) to

$$\sum_{-\infty}^{\infty} (-)^j e^{-2\alpha^2 j^2}.$$

The fact that the two events differ by reason of occurrences such as that indicated by an arrow in fig. 8, clearly cannot affect the probability asymptotically since the height of each jump in the irregular line is only m^{-1} while the width of the region is of order $m^{-\frac{1}{2}}$. Thus

$$P\left\{ \max_p \sqrt{m} \left| \frac{\sum_1^p I_n(\psi_j, \varepsilon)}{\sum_1^m I_n(\psi_j, \varepsilon)} - \frac{p}{m} \right| < \alpha \right\} \rightarrow \sum_{-\infty}^{\infty} (-)^j e^{-2\alpha^2 j^2}. \quad (9)$$

This last function is tabulated in Smirnov [1],[1] the values 1·36 and 1·63 of α corresponding to probabilities 0·05 and 0·01 respectively.

[1] The introductory note to this paper incorrectly states the function being tabulated, a factor 2 being omitted in the exponent.

It is not very difficult to show that we may replace $I_n(\psi_j, \varepsilon)$ in (9) by $I_n(\psi_j, x)/(2\pi f(\psi_j))$, but in general the replacement of $f(\psi_j)$ by $\hat{f}(\psi_j)$ will not be, even asymptotically, valid. This test should also be

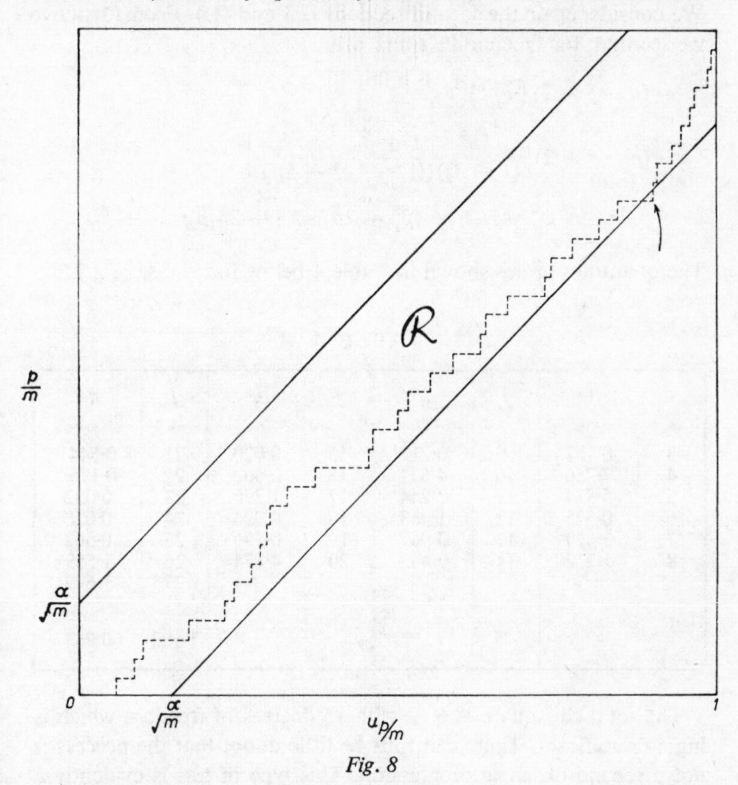

Fig. 8

relatively little affected by departures from normality in the ε_t since the $I_n(\psi_j, \varepsilon)$ are fundamentally constructed from the serial correlations for the ε_t, which as we know have distributions which for large n are substantially independent of the normality assumption. Indeed the variance of the $u_{p/m}$ is independent of κ_4 to the order n^{-1} as is easily checked.

We close this section with an example.

Example

We consider again the data of sections II 3 and II.4. From (3) above we see that, for Quenouille's test,

$$d_j^2 = (n - j + 2)\{\hat{\Phi}_{j,n}^{(2)}\}^2$$

$$= (n - j + 2)\left[\frac{1 + \hat{\beta}_2}{(1 - \hat{\beta}_2)\{(1 + \hat{\beta}_2)^2 - \hat{\beta}_1{}^2\}}\right]$$

$$\times \{r_j + 2\hat{\beta}_1 r_{j-1} + (\hat{\beta}_1^2 + 2\hat{\beta}_2)r_{j-2} + 2\hat{\beta}_1\hat{\beta}_2 r_{j-3} + \hat{\beta}_2^2 r_{j-4}\}^2$$

The quantities d_j^2 are shown in Table 4 below for $j = 3, \ldots, 27$.

TABLE 4

j	d_j^2	j	d_j^2	j	d_j^2	j	d_j^2
3	0·737	9	0·712	15	0·020	21	0·544
4	7·256	10	4·651	16	0·002	22	0·175
5	5·834	11	7·244	17	1·296	23	0·032
6	0·525	12	1·383	18	0·924	24	0·028
7	3·429	13	0·047	19	0·296	25	0·562
8	1·212	14	9·432	20	4·734	26	1·559
						27	1·649
						Total	60·283

The total chi-square is 60·3 with 25 degrees of freedom which is highly significant. There can thus be little doubt that the process is not a second order autoregression. This type of test is evidently a fairly searching one since an autoregression of order 27 (the range of alternatives considered) could be used to graduate a very wide range of processes.

Bartlett [3] shows the complete periodogram for this process (see also fig. 4 above). He grouped the $\sigma^2 I_n(\psi_j)\{2\pi f(\psi_j)\}^{-1} = y_j^2$ as shown in Table 5 below, on prior grounds.

TABLE 5

j	y_j^2	$2n_j$	j	y_j^2	$2n_j$
1·2	0·14013	4	12	0·41206	2
3·4	0·40526	4	13	0·01589	2
5·6	0·05908	4	14	0·00179	2
7	0·02310	2	15, 16	0·02989	4
8	0·07789	2	17, 18, 19	0·04600	6
9	0·01371	2	20, . . ., 28	1·15438	18
10	0·05135	2	29, . . ., 57	4·87392	57
11	0·01707	2			
			Total	7·32154	113

The formula (8) becomes

$$M = K37{\cdot}70$$

$$K = \left[1 + \frac{1}{3(15-1)}\{5{\cdot}21513\}\right]^{-1}.$$

Thus $M = 33{\cdot}5$, which as chi-square with 14 degrees of freedom is highly significant.

Finally the most extreme value of

$$\left[\sum_1^p I_n(\psi_j)/\{2\pi\hat{f}(\psi_j)\}\right] \Bigg/ \left[\sum_1^m I_n(\psi_j)/\{2\pi\hat{f}(\psi_j)\}\right]$$

occurs at $p = 30$ where its difference from 30/56 is $-0{\cdot}184$ so that the maximum value of the test statistic in (9) is 1·39. This just exceeds the 5% point. In the present case this test seems decidedly weaker than the other two and in part at least, as pointed out by Bartlett, this arises from the necessity of weighting the $I_n(\psi_j)$ by dividing by $\hat{f}(\psi_j)$. This results in the part of the spectrum having the greatest total content (i.e. corresponding to the greatest $f(\psi_j)$) being given the least weight while from the point of view of the power of a test it is the most interesting.

It is of course not very practical to test the conditions for the validity of any of these tests and in particular the condition that $\alpha_j = o(j^{-3/2})$ for the third. In the case of the second test $n^{-1}n_j$ is small save for one or two groups out of the 15 so that non-normality of the ε_t should not matter greatly. Quenouille's test, for the present case of an autoregressive hypothesis, because of its properties of robustness, computational simplicity and power appeals as the most useful of the three.

4. Confidence intervals

The result (9) of the last section may be used, under the conditions of the theorem, to give (asymptotically) a confidence interval for $f(\lambda)$ (or more properly for $f(\lambda)$ at the points ψ_j) by solving for y_j the equations (for the upper limit)

$$\sum_1^p y_j \begin{cases} = \left(\frac{\alpha}{\sqrt{m}} + \frac{p}{m}\right) \sum_1^m y & 1 \leqslant p \leqslant (m - \sqrt{m}\,\alpha) \\ = 1 & (m - \sqrt{m}\,\alpha) < p \end{cases}$$

and then putting $f_u(\psi_j) = y_j^{-1} I_n(\psi_j)$ where f_u is the upper limit. A similar set of equations would be used for the lower limit. These equations are not easy to solve, however, so that the method is of more theoretic than practical interest. Here we propose to discuss an alternative method of obtaining a confidence interval for $f(\lambda)$ due to Grenander and Rosenblatt.[1] The Grenander and Rosenblatt method, though of considerable mathematical interest, is of reduced practical value because of its present dependence upon the assumption of normality. We shall therefore discuss it in an abbreviated and heuristic form (also due to the authors of the method).[2] The method is discussed in detail, with admirable clarity and precision, in Grenander and Rosenblatt [1], [3].

Grenander and Rosenblatt [1, 2] consider the case of a linear process x_t for which ε_t has $\kappa_4 = 0$, and $\alpha_j = o(j^{-3/2})$. They form the statistics

$$G_n^*(\lambda) = \frac{1}{4\pi} \int_0^\lambda I_n(\theta)\, d\theta$$

[1] For other suggestions relating to a confidence interval for $f(\lambda)$ see Bartlett [3] and Grenander and Rosenblatt [1] Chapter 6 especially section VI.11.

[2] Grenander and Rosenblatt [2].

whose covariance properties are obtainable from formula (III.2.12) as[1]

$$n \operatorname{cov} \{G_n^*(\lambda)G_n^*(\mu)\} = 2\pi \int_0^{\min \lambda, \mu} f^2(\theta)\, d\theta.$$

Consider

$$y(\lambda) = \sqrt{2\pi} \int_0^{\lambda} f(t)\, dz(t)$$

where $z(t) - z(\tau)$ is normally distributed independently of $z(\tau)$, $t > \tau$, and $\mathscr{E}\{z(t)z(\tau)\} = \min t, \tau$. Then

$$\mathscr{E}\{y(\lambda)y(\mu)\} = 2\pi \int_0^{\min \lambda, \mu} f^2(\theta)\, d\theta$$

and we might expect that

$$\lim_{n\to\infty} P\left\{ \max_{0\leqslant\lambda\leqslant\pi} \sqrt{n} \left| G_n^*(\lambda) - \int_0^{\lambda} f(\theta)\, d\theta \right| < \alpha \right\}$$
$$= P\left\{ \max_{0\leqslant\lambda\leqslant\pi} \sqrt{2\pi} \left| \int_0^{\lambda} f(t)\, dz(t) \right| < \alpha \right\} \quad (1)$$

since the increments in $G_n^*(\lambda)$ could also be expected to be normal for n large and the two processes have, asymptotically, the same covariance function.

Now $y(\lambda)$ is also a process of independent (normal) increments like $z(t)$ but with a change of scale in the variable:

$$t = 2\pi \int_0^{\lambda} f^2(\theta)\, d\theta.$$

In determining probabilities relating to the maximum value of a process such as $y(\lambda)$, over a certain interval, the transformation of the time scale clearly will not matter. All that will matter will be the length of the interval so that we normalize $y(\lambda)$ by dividing by $\left\{2\pi \int_0^{\pi} f^2(\lambda)\, d\lambda\right\}^{1/2}$ to obtain a process which, after a change to a

[1] The result from formula (III.2.12) must be halved since the definition of $G_n^*(\lambda)$ does not include $I_n(-\theta)$ as well as $I_n(\theta)$.

new scale, will correspond to a process of the form of $z(t)$, with t running over an interval of unit length.

Then

$$P\left\{ \max_{0 \leq \lambda \leq \pi} \left| \frac{\sqrt{2\pi} \int_0^\lambda f(t)\, dz(t)}{\left\{ 2\pi \int_0^\pi f^2(\lambda)\, d\lambda \right\}^{\frac{1}{2}}} \right| < \alpha \right\} = P\left\{ \max_{0 \leq t \leq 1} \left| z(t) \right| < \alpha \right\}$$

should equal in the limit

$$P\left\{ \max_{0 \leq \lambda \leq \pi} \sqrt{n} \left| G_n^*(\lambda) - \int_0^\lambda f(\theta)\, d\theta \right| < \alpha H \right\} \tag{2}$$

where we have replaced $2\pi \int_0^\pi f^2(\lambda)\, d\lambda$ by H, in (2), where

$$H = \tfrac{1}{2} c_0^2 + \sum_1^m c_j^2, \quad m = O(n^\delta), \quad 0 < \delta < 1$$

which is consistent. Since (Sommerfeld [1] p. 71)

$$P\left\{ \max_{0 \leq t \leq 1} | z(t) | < \alpha \right\} = \sum_{-\infty}^{\infty} (-)^j [\Phi\{(2j+1)\alpha\} - \Phi\{(2j-1)\alpha\}] \tag{3}$$

where $\Phi(x) = (2\pi)^{-\frac{1}{2}} \int_{-\infty}^{x} e^{-\frac{1}{2}y^2}\, dy$

the asymptotically known probability for (2) enables a confidence interval for $F(\lambda)$ to be constructed.

If $\kappa_4 \neq 0$ the probability of the left-hand expression in (1) may again be, asymptotically, expressed in terms of a probability relating to a normal process $y(\lambda)$ with zero mean but now the covariance becomes

$$\mathscr{E}\{y(\lambda)y(\mu)\} = \kappa_4 \int_0^\lambda f(\theta)\, d\theta \int^\mu f(\theta)\, d\theta + 2\pi \int^{\min \lambda,\, \mu} f^2(\theta)\, d\theta$$

and the probability relating to the $y(\lambda)$ process is much more difficult to determine and depends upon $f(\lambda)$ in an essential way. The case where $f(\lambda)$ is uniform has been treated and the values of α for probabilities 0·9, 0·95, 0·99 and 0·999 are given in Grenander and Rosenblatt [1]. If κ_4 is not known it may be estimated consistently and the estimate used in place of κ_4 but, apparently, an estimate which converges rapidly is not easy to find.

Clearly (1) will also give a test of goodness of fit of a prescribed $f(\lambda)$. Aside from the difficulty due to κ_4 this test is also deficient insofar as $f(\lambda)$ must be completely prescribed (apart from a multiplicative constant) since $\max_{\lambda} \sqrt{n} \left| \int_0^{\lambda} (f(\theta) - \hat{f}(\theta))\, d\theta \right|$ will not in general converge in probability to zero. This last difficulty disappears when the problem is that of testing whether two series $x_{1,t}$ and $x_{2,t}$ (of n_1 and n_2 observations) are generated by the same process for then the statistic

$$\max_{0 \leqslant \lambda \leqslant \pi} \left| \frac{\sqrt{n}\{G_{n_1}{}^*(\lambda) - G_{n_2}{}^*(\lambda)\}}{\left\{H_{n_1}{}^2 + H_{n_2}{}^2\right\}^{\frac{1}{2}}} \right| ; \quad n = \frac{2n_1 n_2}{n_1 + n_2}$$

has, when n_1 and n_2 increase indefinitely so that $n_2^{-1} n_1 \longrightarrow c > 0$, the same asymptotic distribution as that given by (3) above. Here the n_1 and n_2 suffices indicate which sets of observations the relevant statistics are computed from. The difficulty due to the fact that κ_4 will not be zero remains, however.

CHAPTER V

Processes Containing a Deterministic Component

1. Introduction

Throughout the first four chapters we have considered processes $\{x_t\}$ having mean value zero, though in many cases, where it was more convenient to do so in that place, we have made the mean correction. In this chapter we proceed to consider the problem of a process $\{x_t\}$ having a non-zero mean, possibly dependent upon t. As such this chapter can be looked upon as an attempt to extend the results of the first four chapters to a class of processes, containing a deterministic and possibly evolutive component, which may be reduced to stationary form by the simple subtraction of a time dependent mean. Such a preliminary treatment of data will nearly always be necessary before the methods of the previous three chapters can be applied.

We consider therefore the linear regression of a variable z_t upon k 'regressor' variables $y_{1,t}, \ldots, y_{k,t}$. The variable z_t is regarded as being generated by a relation of the form

$$z_t = \sum_{j=1}^{k} \delta_j y_{j,t} + x_t \tag{1}$$

where x_t is generated by a stationary process. Among the $y_{j,t}$ will usually occur, of course, a variable identically equal to one. If n observations are available we shall write (1) in the form

$$\mathbf{z} = \mathbf{Y}\boldsymbol{\delta} + \mathbf{x} = [\mathbf{y}_1\mathbf{y}_2 \ldots \mathbf{y}_k]\boldsymbol{\delta} + \mathbf{x}$$

where the $\mathbf{y}_j$, $\mathbf{z}$ and $\mathbf{x}$ are vectors, of n components, and $\boldsymbol{\delta}$ is a vector of k components. We shall indicate the vector corresponding to the mean correction, for which $y_{j,t} \equiv 1$, by $\mathbf{1}$.

The $y_{j,t}$ will be regarded as fixed numbers. In some cases they will be the values at equidistant points in time of analytic functions. In others, however, they will not be generated by any such deterministic process, but we shall only consider cases where the $y_{j,t}$ are independent of x_t so that the inferences will be made conditionally upon the fixing of the $y_{j,t}$ at their observed values.[1]

We discuss the estimation of the δ_j in the next section. It will be shown that the departure of the process generating x_t from a process of independent random variables may affect both the efficiency of the least square methods and the validity of the usual tests of significance. In the third section we shall consider the problem of testing for such a departure, and in the fourth the particular case of trend removal. In the final section we shall give a short discussion of the effects of the process of estimating the δ_j on the validity of the application of the methods of the previous three chapters to the estimated residuals.

2. The estimation of the regression coefficients

If $\mathbf{\Gamma}_n$ is the covariance matrix of the n observed values of x_t,[2] then the least squares estimate of $\boldsymbol{\delta}$ is

$$\hat{\boldsymbol{\delta}} = [\mathbf{Y}'\mathbf{Y}]^{-1}\mathbf{Y}'\mathbf{z} = \boldsymbol{\delta} + [\mathbf{Y}'\mathbf{Y}]^{-1}\mathbf{Y}'\mathbf{x} \tag{1}$$

so that the covariance matrix of the $\hat{\boldsymbol{\delta}}_j$ is

$$[\mathbf{Y}'\mathbf{Y}]^{-1}\mathbf{Y}'\mathbf{\Gamma}_n\mathbf{Y}[\mathbf{Y}'\mathbf{Y}]^{-1}. \tag{2}$$

It is a classic fact (see for example Aitken [1]) that the unbiased estimator of the vector $\boldsymbol{\delta}$ which is best in the sense that it minimizes the variance of the estimate is that given by

$$\tilde{\boldsymbol{\delta}} = [\mathbf{Y}'\mathbf{\Gamma}_n^{-1}\mathbf{Y}]^{-1}\mathbf{Y}'\mathbf{\Gamma}_n^{-1}\mathbf{z} = \boldsymbol{\delta} + [\mathbf{Y}'\mathbf{\Gamma}_n^{-1}\mathbf{Y}]^{-1}\mathbf{Y}'\mathbf{\Gamma}_n^{-1}\mathbf{x}$$

The covariance matrix of the vector $\tilde{\boldsymbol{\delta}}$ is easily seen to be

$$[\mathbf{Y}'\mathbf{\Gamma}_n^{-1}\mathbf{Y}]^{-1}. \tag{3}$$

[1] The title of this chapter is somewhat misleading as the $y_{j,t}$ could themselves be generated by stationary, purely non-deterministic processes. As this book is confined to univariate processes, however, the chapter will emphasize the case where the $y_{j,t}$ correspond to what would usually be called a trend.

[2] We shall continue to use, for the parameters involved in the prescription of the process generating x_t, the same symbols as have been used in previous chapters.

The form of estimate, $\tilde{\boldsymbol{\delta}}$, is not of much direct use since $\boldsymbol{\Gamma}_n$ is not known. In cases where the nature of the process generating the x_t is prescribed *a priori*, save for certain unspecified parameters, the likelihood equations (on the basis of normality) could be written down and solved. This will usually be a difficult procedure, however. In some cases, in economic applications, it has been argued (from experience) that the residuals will be generated by a process near to a simple Markoff process with high positive ρ and that the use of a transformation of the type

$$z_t \longrightarrow z_t - z_{t-1}$$

$$y_{j,} \longrightarrow y_{j,t} - y_{j,t-1} \quad j = 1, \ldots, k$$

will result in a new relation in the differenced quantities having a set of residuals sufficiently near to independent for the application of (1) with $\boldsymbol{\Gamma}_n = \mathbf{I}_n$. We shall discuss this further below.[1] Failing the possession of any such precise prior information the only reasonable procedure would seem to be the carrying out of an initial regression based on the assumption $\boldsymbol{\Gamma}_n = \mathbf{I}_n$ and the examination of the residuals from this regression by the methods of the last three chapters (see the last section of this chapter also). One would hope, and experience to some extent fortifies us, that the process describing the x_t could be reasonably well described by a low order autoregression. The autoregressive constants having been estimated one would make the transformation

$$z_t \longrightarrow \sum_0^p \hat{\beta}_k z_{t-k}$$

$$y_{j,t} \longrightarrow \sum_{k=0}^p \hat{\beta}_k y_{j,t-k}$$

and proceed to estimate the δ_j from the transformed quantities.

[1] The differencing operation clearly nullifies $\mathbf{1}$ so that the corresponding δ_j (say δ_1) will have to be estimated separately, by $\bar{z} - \sum_j \delta_j{}^* \bar{y}_j$ where $\delta_j{}^*$ is the estimate of δ_j obtained from the differenced variables.

Under reasonable conditions such a procedure should result in fairly efficient estimates for the δ_j. *Asymptotically* such a procedure will certainly be fully efficient, if the residuals are generated by an autoregression.

In order to judge these procedures we have just discussed let us consider the effect of a wrong prescription of the nature of the process generating the x_t. The effect of this upon the efficiency of the least squares procedure will depend upon the nature of the vectors $\mathbf{y}_j$. Indeed if the $\mathbf{y}_j$ are proper vectors of $\mathbf{\Gamma}_n$ the procedure based on putting $\mathbf{\Gamma}_n = \mathbf{I}_n$ in (1) is *numerically* the same as the best linear unbiased (b.l.u.) procedure, as is easily checked. This is a situation which, strictly speaking, never occurs (unless $\mathbf{\Gamma}_n = \mathbf{I}_n$) though we shall see that there are situations which approximate to it.

We need consider only the case where the *assumed* covariance matrix of the x_t is $\gamma_0 \mathbf{I}_n$ since for the case where it is $\gamma_0 \mathbf{A}_n$, say, we shall make a transformation $\mathbf{z} \longrightarrow \mathbf{Qz}$, $\mathbf{y}_j \longrightarrow \mathbf{Qy}_j$ such that

$$\mathbf{Q}'\mathbf{A}_n\mathbf{Q} = \mathbf{I}_n$$

and arrive at a *true* covariance matrix for the new residuals $\mathbf{Q}'\mathbf{\Gamma}_n\mathbf{Q}$ which we rename $\mathbf{\Gamma}_n$.

In the case where $k = 1$ the ratio of the two variances becomes, from (2) and (3) above,

$$\text{eff}\,(\hat{\delta}_1) = \frac{\text{var}\,(\tilde{\delta}_1)}{\text{var}\,(\hat{\delta}_1)} = \frac{(\mathbf{y}_1'\mathbf{y}_1)^2}{(\mathbf{y}_1'\mathbf{\Gamma}_n\mathbf{y}_1)(\mathbf{y}_1'\mathbf{\Gamma}_n^{-1}\mathbf{y}_1)}; \tag{4}$$

a suitable measure of the efficiency of $\hat{\delta}_1$. A (sharp) lower bound to (4) is obtained when $\mathbf{y}_1 = \boldsymbol{\sigma}_1 \pm \boldsymbol{\sigma}_n$ where $\boldsymbol{\sigma}_1$ and $\boldsymbol{\sigma}_n$ are the normalized proper vectors of $\mathbf{\Gamma}_n$ corresponding, respectively, to the least and greatest proper values (the same lower bound being taken for both signs). Thus

$$\left[\frac{1}{2}\left\{\left(\frac{\mu_1}{\mu_n}\right)^{1/2} + \left(\frac{\mu_n}{\mu_1}\right)^{1/2}\right\}\right]^{-2} \leqslant \text{eff}\,(\hat{\delta}_1) \leqslant 1 \tag{5}$$

the upper bound being taken of course when y_1 is any proper vector of $\mathbf{\Gamma}_n$. Here $\mu_1 \leqslant \mu_2 \ldots \leqslant \mu_n$ are the n, ordered proper values of $\mathbf{\Gamma}_n$.

For by a rotation we may reduce the problem to that of minimizing

$$\left(\sum_1^n w_j\right)^2 \Bigg/ \left\{\sum_1^n \mu_j w_j \sum_1^n \mu_j^{-1} w_j\right\}$$

subject to $w_j \geqslant 0$.

We may also presume that no two of the μ_j are equal since otherwise the problem may be reduced to one in fewer dimensions. The theorem is trivially proved for $n = 2$ and for $n > 2$ it will be shown that the minimum is attained when only two of the w_j are not zero. This will establish the result. If a maximum M were attained at a point $w_j = w_{j,0}$ with, say $w_{k,0} \neq 0, k = 1, 2, 3$ then

$$\left[\frac{\partial}{\partial w_k}\left\{M^{-1}\left(\sum_1^n w_j\right)^2 - \sum_1^n \mu_j w \sum_1^n \mu_j^{-1} w_j\right\}\right]_{w_j = w_{j,0}} = 0;$$

$$k = 1, 2, 3$$

so that

$$2M^{-1} \sum_1^n w_{j,0} - \mu_k \sum_1^n \mu_j^{-1} w_{j,0} - \mu_k^{-1} \sum_1^n \mu_j w_j = 0;$$

$$k = 1, 2, 3.$$

But, as a set of equations in $\sum w_{j,0}$, $\sum \mu_j w_{j,0}$ and $\sum \mu^{-1} w_{j,0}$, this system has the determinant

$$-\frac{2M^{-1}}{\mu_1\mu_2\mu_3}(\mu_1 - \mu_2)(\mu_1 - \mu_3)(\mu_2 - \mu_3) \neq 0$$

so that, since $\sum w_{j,0}$, $\sum \mu_j w_{j,0}$ and $\sum \mu_j^{-1} w_{j,0}$ are certainly not zero at the maximum, a contradiction is reached.

A precise treatment for $k > 1$ can also be given[1] but for our purposes it will be sufficient to observe[2] that if $\mathbf{\Gamma}_n$ corresponds to a process with spectral density $f(\lambda)$ it follows from the results of section I.5 that the μ_j will be, approximately, the values of $f(\lambda)$ at the

[1] See Watson [3]. The proof there given is faulty but Dr Watson in a personal communication has provided a valid proof. [2] See Watson and Hannan [1].

points ψ_j. If n is reasonably large, $f(\lambda)$ is not too irregular, and k is small relative to n, then

$$\mu_1 \approx \mu_2 \ldots \approx \mu_k,$$
$$\mu_{n-k+1} \approx \mu_{n-k+2} \approx \ldots \approx \mu_n.$$

The left-hand side of (5) will then be a lower bound *almost* simultaneously attainable for all k of the $\hat{\delta}_j$. This may be seen by taking the $\mathbf{y}_j$ to be the set of mutually orthogonal vectors of the form

$$\boldsymbol{\sigma}_j \pm \boldsymbol{\sigma}_{n-j+1} \quad j = 1, \ldots, k$$

(one of the two signs being chosen, it does not matter which). Now

$$[\mathbf{y}_1, \ldots, \mathbf{y}_k]' \boldsymbol{\Gamma}_n^{-1} [\mathbf{y}_1, \ldots, \mathbf{y}_k]$$

is diagonal with $\left(\mu_j^{-1} + \mu_{n-j+1}^{-1}\right)$ in the jth place in the principal diagonal and zeros elsewhere. Also

$$[\mathbf{Y}'\mathbf{Y}]^{-1}\mathbf{Y}'\boldsymbol{\Gamma}_n\mathbf{Y}[\mathbf{Y}'\mathbf{Y}]^{-1}$$

has $\frac{1}{4}(\mu_j + \mu_{n-j+1})$ in the jth place in the main diagonal and zeros elsewhere. Thus, for these $\mathbf{y}_j$,

$$\text{eff}\,(\hat{\delta}_j) = \left[\frac{1}{2}\left\{\left(\frac{\mu_j}{\mu_{n-j+1}}\right)^{\frac{1}{2}} + \left(\frac{\mu_{n-j+1}}{\mu_j}\right)^{\frac{1}{2}}\right\}\right]^{-2} \approx \text{eff}\,(\hat{\delta}_1)$$

and the left-hand side of (5) may be used as a measure of the lower bound to the efficiency of the least squares procedure, for given $\boldsymbol{\Gamma}_n$. This lower bound is seen to depend upon the ratio of the least and greatest values of $f(\lambda)$.

One must emphasize, of course, that the lower bound will in general not be attained so that the expression (5), for given $f(\lambda)$, may be far too pessimistic for the observed $y_{j,t}$.

Example 1

Consider the case where x_t is generated by

$$x_t + \beta_1 x_{t-1} + \beta_2 x_{t-2} = \varepsilon_t \qquad (6)$$

while it is assumed that it is of the form

$$x_t - \rho x_{t-1} = \varepsilon_t$$

with ρ equal to the first serial correlation resulting from (6). Thus the case we are considering is asymptotically equivalent to the case where a second order autoregression is presumed to be a Markoff process with ρ estimated from the data to hand. The lower bound, for $k = 1$, to the efficiency is then $(1 + \beta_2^2)^{-2}(1 - \beta_2^2)^2$ which is shown for certain β_2 in Table 6 below.

TABLE 6

β_2	0	0·1	0·2	0·3	0·4	0·5	0·6	0·7	0·8	0·9
$\left(\frac{1-\beta_2^2}{1+\beta_2^2}\right)^2$	1·00	0·96	0·85	0·70	0·52	0·36	0·22	0·12	0·05	0·01

Example 2

If x_t is generated by

$$x_t - \rho x_{t-1} = \varepsilon_t, \quad \rho \leqslant 1 \tag{7}$$

and a first difference transformation is used, the spectral density of $(x_t - x_{t-1})$ becomes proportional to

$$(1 - \cos \lambda)/(1 + \rho^2 - 2\rho \cos \lambda)$$

whose minimum value is zero, for $\lambda = 0$, if $\rho < 1$. Thus the *lower bound* to the efficiency of the least squares estimate from the transformed equation is now zero, asymptotically, and will be near to zero even for quite small n. Of course the regressor variables met in practice will, most often, *not* be such that the differenced variables will be of the form which minimizes the efficiency of the least squares estimates of the δ_j.

However, if the regression relation is

$$z_t = \delta_1 + \delta_2 t + x_t \tag{8}$$

the differenced relation becomes

$$(z_t - z_{t-1}) = \delta_2 + (x_t - x_{t-1})$$

and the least squares estimate is

$$\delta_2^* = (z_n - z_1)/n$$

having variance $2n^{-2}(1 - \rho^{n-1})$.

The b.l.u. estimate (based on a knowledge of ρ) will, effectively, be obtained by returning to the original equation, (8), and making the autoregressive transformation. This will result in an estimate

$$\tilde{\delta} = \frac{\sum_{2}^{n} \left\{ (t - \rho(t-1)) - \frac{n-1}{2}(1-\rho) - \frac{n-1}{n} \right\} \{z_t - \rho z_{t-1}\}}{\sum_{2}^{n} \left\{ (t - \rho(t-1)) - \frac{n-1}{n}(1-\rho) - \frac{n-1}{n} \right\}^2}$$

whose variance is clearly $O(n^{-3})$.

Thus the lower bound to the efficiency is attained, at least asymptotically, in this eminently important case. As we shall see below the application of straightforward least squares to the *original* relation results in an estimate of δ_2 (and δ_1) which will be *asymptotically* efficient. A similar situation arises in the general case of a polynomial trend. When polynomial trends are being eliminated; and n is large, differencing should not be carried out (except in the unlikely situation where it is known that ρ in (7) is unity).

On the other hand when $k = 1$ and $y_{j,t}$ is also a simple Markoff process, with parameter ρ_1, the estimate of δ_1 from the first differences has efficiency (asymptotically) which is

$$\frac{2(1-\rho_1)(1-\rho\rho_1)(1+\rho)}{(3-\rho-\rho_1-\rho\rho_1)(1+\rho^2-2\rho\rho_1)}.$$

For example for $\rho = 0{\cdot}9$ and $\rho_1 = 0{\cdot}8$ this is 0·34. The efficiency of the straightforward least squares procedure is here (asymptotically)

$$\frac{(1-\rho^2)(1-\rho\rho_1)}{(1+\rho\rho_1)(1+\rho^2-2\rho\rho_1)}$$

which is 0·03 for these values of ρ and ρ_1.

The general nature of the situation will be further investigated in section 4.

The fact that the process generating x_t need not be one of independent random variables also affects the tests of significance for the coefficients δ_j of course. Indeed as we have seen the covariance matrix of these estimates will be $[\mathbf{Y}'\mathbf{Y}]^{-1}\mathbf{Y}'\mathbf{\Gamma}_n\mathbf{Y}[\mathbf{Y}'\mathbf{Y}]^{-1}$ instead of $[\mathbf{Y}'\mathbf{Y}]^{-1}$.

Failing any precise information concerning the nature of the process generating the x_t the only course available would apparently be to estimate the nature of the process, using the residuals from a preliminary regression (based upon $\mathbf{\Gamma}_n = \mathbf{I}_n$) and after a transformation based upon this estimation to proceed to test the significance of the δ on the basis of the large sample theory which would apply if the nature of the process, and hence the appropriate transformation, had been prescribed precisely and not estimated.[1]

As we shall see in the following sections another class of cases arises where straightforward least squares is efficient and therefore will be used, even when the spectrum of x_t is not uniform. (As we have seen, this is so when the $\mathbf{y}_j$ are proper vectors of $\mathbf{\Gamma}_n$.) The tests of significance of the estimated regression coefficients still need modification, however, and we shall also discuss this below.

Example 3

Two series (x_t and y_t), of 70 observations each, were chosen from series 7 of Kendall [3], the starting terms having numbers 191 and 316. The two series are thus simple (Gaussian) Markoff processes with parameter 0·9. They are sufficiently far apart to be treated as effectively independent. A third series, z_t, was formed by

$$z_t = 0{\cdot}8 y_t + x_t$$

The corrected sums of squares and cross products for the regression of z_t on y_t are

$$\begin{array}{cc} z & y \\ \end{array}$$
$$\begin{bmatrix} 274{\cdot}698 & 65{\cdot}716 \\ & 157{\cdot}530 \end{bmatrix}$$

The estimate $\hat{\delta}$ is thus 0·417. If the data is treated as if x_t were generated by a process of *independent* Gaussian variables then the 5% confidence interval for δ is

$$0{\cdot}113 \leqslant \delta \leqslant 0{\cdot}721$$

which does not include the true value. Of course this treatment

[1] If x_t can be prescribed to be a normal autoregressive process (with unprescribed autoregressive constants) exact tests procedures are available (which will not be very sensitive to departures from normality). See Hannan [1]. However, these exact tests are inefficient.

greatly underestimates the true variance of $\hat{\delta}$. In fact, it may be seen from (V.2.1) that the variance of $\hat{\delta}$ is, approximately,

$$\frac{1}{n}\frac{\sigma_x^2}{\sigma_y^2}\left(\frac{1+\rho^2}{1-\rho^2}\right) = 0{\cdot}083$$

so that the observed value, 0·417, is not significantly different from 0·800.

The observed first serial correlation of the residuals is 0·779 while the first partial serial correlation is 0·008. We therefore make the autoregressive transformation based upon the estimate $\hat{\rho} = 0{\cdot}8$, rounding to a convenient value. The estimate of δ from the transformed data is 0·869, a great improvement.

In the present instance first differencing the series will be a reasonable procedure. The estimate of δ from the differenced observations is 0·907, which is again an improvement on the straightforward least squares estimate.

Example 4

The regression of the 480 values of Kendall [1] series 1 on t gives an estimate for the regression coefficient of 0·024 (the true value being zero of course). The estimate from the differenced observations is 0·079. Although the asymptotic efficiency of the second estimate is zero its variance is still $O(n^{-2})$ so that it is in no way surprising that the estimate, from 480 observations, is quite accurate.

3. Testing the $\mathbf{x}_t$ for departure from independence

Since the nature of the process generating the x_t has a profound effect upon the efficiency of least squares procedure it will often be necessary to test whether the x_t are mutually independent. In large samples this could be done by using the elements of

$$\hat{\mathbf{x}} = [\mathbf{I} - \mathbf{Y}[\mathbf{Y}'\mathbf{Y}]^{-1}\mathbf{Y}']\mathbf{z} = \mathbf{Q}\mathbf{z} \tag{1}$$

as if they were the unobservable x_t, making use of the methods discussed in earlier chapters. We shall discuss this in the last section of this chapter. Here we wish to discuss a small sample test of a more precisely defined null hypothesis. We now require that, on this null hypothesis, the x_t are independent and normal with zero mean and variance γ_0. We consider a test statistic of the form

$$r = \hat{\mathbf{x}}'\mathbf{B}\hat{\mathbf{x}}/\hat{\mathbf{x}}'\hat{\mathbf{x}}. \tag{2}$$

Here **B**, a symmetric matrix, will be chosen so that r is some form of serial correlation. Naturally no single choice of **B** can lead to a test of uniformly high power against *all* alternatives and a reasonable procedure will be to choose **B** so that r is (effectively) the first serial correlation of the elements of $\hat{\mathbf{x}}$. One will be unlikely to strike a process generating x_t having $\rho_1 = 0$ (unless the x_t *are* independent) so that such a test statistic should have fair power against a wide range of alternatives.

Thus, since $\mathbf{Qz} = \mathbf{Qx}$ and **Q** is idempotent of rank $n - k$ we have

$$r = \mathbf{x}'\mathbf{QBQx}/\mathbf{x}'\mathbf{Qx}.$$

Since **Q** and **QBQ** commute and are symmetric we obtain

$$r = \sum_{1}^{n-k} v_j \xi_j^2 \bigg/ \sum_{1}^{n-k} \xi_j^2; \quad v_1 \leqslant v_2 \leqslant \ldots \leqslant v_{n-k}$$

where the v_j are the proper values of **QBQ** other than k zeros. The ξ_j are again normal and independent and may clearly be taken to have unit variance.

Q is a projection on to the orthogonal complement of the subspace $\mathcal{M}$ spanned by the column vectors of **Y**. In general in this subspace, $\mathcal{M}$, there will lie proper vectors of **B** (as we shall see below **1** will be a proper vector of a suitable **B**). Let s be the number of these and let them span the subspace $\mathcal{M}_1$. Then if $\mathcal{M}_2$ is the orthogonal complement of $\mathcal{M}_1$ in $\mathcal{M}$ we may write

$$\mathbf{Q} = \mathbf{Q}_2\mathbf{Q}_1 = \mathbf{Q}_1\mathbf{Q}_2$$

where $\mathbf{Q}_1$ projects on to the orthogonal complement of $\mathcal{M}_1$ while $\mathbf{Q}_2$ projects on to the orthogonal complement of $\mathcal{M}_2$. Thus

$$\mathbf{QBQ} = \mathbf{Q}_2\mathbf{B}_1\mathbf{Q}_2$$

where $\mathbf{B}_1 = \mathbf{Q}_1\mathbf{B}\mathbf{Q}_1$ is a matrix of rank $n - s$ having the same proper values as **B** save that the s proper values corresponding to the s proper vectors spanning $\mathcal{M}_1$ have been removed. Let the remaining proper values of $\mathbf{B}$ (i.e. those of $\mathbf{B}_1$) be

$$\lambda_1 \leqslant \lambda_2 \leqslant \ldots \leqslant \lambda_{n-s}.$$

Now the quadratic form with $\mathbf{Q}_2\mathbf{B}_1\mathbf{Q}_2$ as matrix may alternatively be regarded as a quadratic form with $\mathbf{B}_1$ as matrix formed from a vector which is constrained to lie in the subspace upon which $\mathbf{Q}_2$ projects, i.e. is subjected to $k - s$ linear constraints. It follows from Courant's maxim–minimum principle[1] that

$$\lambda_j \leqslant \nu_j \leqslant \lambda_{j+k-s}.$$

Thus we have the inequality

$$\frac{\sum_1^{n-k} \lambda_j \xi_j^2}{\sum_1^{n-k} \xi_j^2} \leqslant r = \frac{\sum_1^{n-k} \nu_j \xi_j^2}{\sum_1^{n-k} \xi_j^2} \leqslant \frac{\sum_1^{n-k} \lambda_{j+k-s} \xi_j^2}{\sum_1^{n-k} \xi_j^2}. \tag{3}$$

The two bounding ratios are now in no way dependent upon the vectors in $\mathbf{Y}$ and for a $\mathbf{B}$ of fixed form and given n, k and s their distribution can be established once and for all. The distribution problem is precisely that discussed in section IV.2 (*a*). Durbin and Watson [1], to whom this work is due, chose $\mathbf{B}$ of the form

$$\mathbf{B}_d = \begin{bmatrix} 1 & -1 & 0 & \ldots & 0 & 0 \\ -1 & 2 & -1 & \ldots & 0 & 0 \\ \cdot & \cdot & \cdot & \cdot & \cdot & \cdot \\ \cdot & \cdot & \cdot & \cdot & \cdot & \cdot \\ \cdot & \cdot & \cdot & \cdot & \cdot & \cdot \\ 0 & 0 & 0 & & 2 & -1 \\ 0 & 0 & 0 & & -1 & 1 \end{bmatrix}$$

which makes

$$\hat{\mathbf{x}}'\mathbf{B}_d\hat{\mathbf{x}} = \sum_2^n (\hat{x}_t - \hat{x}_{t-1})^2.$$

The proper values are $\lambda_{n,j} = 2(1 - \cos \tfrac{1}{2}\psi_j)$; $j = 0, \ldots, n-1$; and $\mathbf{1}$ corresponds as proper vector to $\lambda_{n,0} = 0$. Thus putting $s = 1$

[1] See Courant and Hilbert [1] p. 33.

they tabulated the bounds to the 5%, 2·5% and 1% points of r_d for $k' = k - 1 = 1(1)5$; $n = 15(1)40(5)100$, for a one-sided test against positive serial correlation. If $n > 100$ the use of the first serial correlation of the residuals as if no regression had been carried out would suffice.

The only other choice of **B** which has been used is that which makes r the first *circular* serial correlation of the residuals. The matrix of the numerator form we indicated by $\mathbf{W}_1$ in section IV.2. Since $\mathbf{W}_1 = \frac{1}{2}[\mathbf{U} + \mathbf{U}']$ where **U** is the matrix of section I.2 we know that its proper values and proper vectors are

Proper value	*Proper vector(s)*
1	**1**
$\cos \psi_j,\ j=1 \ldots \left[\frac{n-1}{2}\right]$	$\{\cos t\psi_j,\ t=1,\ldots,n\}$, $\{\sin t\psi_j,\ t=1,\ldots,n\}$
for n odd, -1	$\{\cos \pi t,\ t = 1, \ldots, n\}$

The reason for this choice of **B** is now evident since for the case where the mean value of z_t is of the form

$$\alpha_0 + \sum_{1}^{\frac{k-1}{2}} \alpha_j \cos t\psi_j + \sum_{1}^{\frac{k-1}{2}} \beta_j \sin t\psi_j \tag{4}$$

the regressor vectors are proper vectors of $\mathbf{W}_1$. The significance points for a one-sided test (against positive serial correlation in x_t) at the 5% and 1% level are tabulated in Anderson and Anderson [1] for various choices of the terms to be included in (4) and n up to 50 (and higher in some cases). The tests in the cases covered by these tables are of course exact and involve no use of bounds. The essential point about the case of a trigonometric regression is the fact that **B** may be chosen so as to make the regressor vectors proper vectors but so that it is still of a form which should give reasonable power against the type of alternative met with in practice.

Example

We use the data of the example 3 of the last section. The observed first serial correlation of the residuals was 0·779. Since $r_d \approx 2(1 - r_1)$

this statistic has a value near 0·442. In fact the true value is 0·383. The bounds for the significance point for one-sided test against positive serial correlation at the 1% level are 1·43 and 1·49 (entering Durbin and Watson's table with $k - 1 = k' = 1$ and $n = 70$). Thus the observed value is certainly significant since it is below the lower bound.

When an observed value falls between the bounds the test is inconclusive. In such a case the significance point can be located approximately by calculating the true mean and variance of r_d and using the appropriate significance point obtained from a Beta distribution with this mean and variance.

The moments[1] of r_d may be found fairly easily when it is recognized that the ratio is distributed independently of its denominator (Pitman [1]). For, writing $r_d = uv^{-1}$ we then obtain

$$\mathscr{E}(u^s) = \mathscr{E}(r_d^s v_s) = \mathscr{E}(r_d^s)\mathscr{E}(v^s)$$

i.e.

$$\mathscr{E}(r_d^s) = \mathscr{E}(u^s)\{\mathscr{E}(v^s)\}^{-1}.$$

The denominator, v, is a chi-square with $n - k$ degrees of freedom. The numerator is $\sum_1^{n-k} \nu_j \xi_j^2$ and its sth cumulant is thus the sum of the sth cumulants of $\nu_j \xi_j^2$, i.e. (writing Tr (**A**) for the trace of **A**) $\kappa_s(u) = 2^{s-1}(s-1)! \sum_1^{n-k} \nu_i^s = 2^{s-1}(s-1)! \operatorname{Tr} \{[\mathbf{QB}_d\mathbf{Q}]^s\}$. From these results it may be shown that

$$\mathscr{E}(r_d) = \frac{P}{n-k}, \quad \operatorname{var}(r_d) = \frac{2}{(n-k)(n-k-1)}\{Q - P\mathscr{E}(r_d)\}$$

where

$$P = \operatorname{Tr}(\mathbf{B}_d) - \operatorname{Tr}\{\mathbf{Y}'\mathbf{B}_d\mathbf{Y}[\mathbf{Y}'\mathbf{Y}]^{-1}\}$$

$$Q = \operatorname{Tr}(\mathbf{B}_d^2) - 2\operatorname{Tr}\{\mathbf{Y}'\mathbf{B}_d^2\mathbf{Y}(\mathbf{Y}'\mathbf{Y})^{-1}\} + \operatorname{Tr}[\{\mathbf{Y}'\mathbf{B}_d\mathbf{Y}[\mathbf{Y}'\mathbf{Y}]^{-1}\}^2].$$

[1] For **B** of the form which makes r the first circular serial correlation of the $\hat{x}_t$, the mean and variance are derived in Moran [2]. See also Hannan [1] for an alternative test which is exact and asymptotically fully efficient but which may require greatly increased computations.

4. The removal of a trend by regression

(*a*) The efficiency of the least squares estimates.

In this section we consider the case of a process whose mean value is described by some smooth function such as a polynomial in t or a trigonometric polynomial. However, we shall first derive some general results due to Grenander [2][1] which throw light also upon the previous sections and the next section. We consider situations where the regressor variables are generated by some process (not necessarily stochastic) such that

$$\text{(i)}\quad \lim_{n\to\infty} \sum_{1}^{n} y_{j,t}^2 = \lim Y_{n,j}^2 = \infty$$

$$\text{(ii)}\quad \lim_{n\to\infty} \frac{Y_{n+1,j}}{Y_{n,j}} = 1$$

$$\text{(iii)}\quad \lim_{n\to\infty} \frac{\sum_{1}^{n} y_{i,\,t+h} y_{j,t}}{Y_{n,i} Y_{n,j}} = \lim_{n} r_{ij}(h) = \rho_{ij}(h), \quad \text{exists.}$$

The first condition is necessary in order that a consistent estimator of δ should exist. The second ensures that the alteration of 'end terms' in the sequence defining $\rho_{ij}(h)$ will not affect the limit. For example, (ii) combined with (i) shows that

$$\sum_{n-h+1}^{n} y_{j,t+h} y_{i,t} / (Y_{n,i} Y_{n.j}) \to 0$$

so that

$$\rho_{i,j}(-h) = \lim_{n} \frac{\sum_{1}^{n} y_{i,t-h} y_{j,t}}{Y_{n,i} Y_{n,j}} = \lim_{n} \frac{\sum_{1}^{n} y_{j,t+h} y_{i,t}}{Y_{n,i} Y_{n,j}} = \rho_{j,i}(h).$$

If we indicate the $(k \times k)$ matrix having $\rho_{ij}(h)$ in the ith row and jth column by $\mathbf{R}(h)$ then we see that $\mathbf{R}(h) = \mathbf{R}(-h)'$. If we form, with

[1] See also Grenander and Rosenblatt [1].

the vector u having components u_i, the Hermitian form[1]

$$\sigma_{\mu-\nu} = \mathbf{u}^*\mathbf{R}(\mu-\nu)\mathbf{u} = \lim_n \sum_{i,j}^{k} u_i u_j \{Y_{n,i} Y_{n,j}\}^{-\frac{1}{2}} \sum_{t=1}^{n} y_{i,t+\mu} y_{j,t+\nu}$$

(using (i) and (ii) once more) then we see that the sequence σ_t is a positive sequence (see section I.3). For, with any set ξ_j of complex numbers we have

$$\sum_{\mu}^{m}\sum_{\nu=1} \xi_\mu \xi_\nu \sigma_{\mu-\nu} = \lim_n \sum_{1}^{n} \left| \sum_i \left\{ \sum_\mu \bar{u}_i \xi_\mu y_{i,t+\mu} / \sqrt{Y_{n,i}} \right\} \right| \geqslant 0.$$

Thus from the theorem of section I.2 we see that

$$\sigma_t = \int_{-\pi}^{\pi} e^{it\lambda}\, dG_u(\lambda)$$

where $G_u(\lambda)$ is non-decreasing and of total variation σ_0. Since $\mathbf{u}$ is arbitrary we see that this may be written in the form

$$\sigma_t = \mathbf{u}^*\mathbf{R}(t)\mathbf{u} = \int_{-\pi}^{\pi} e^{it\lambda}\, d\{\mathbf{u}^*\mathbf{M}(\lambda)\mathbf{u}\}$$

where $\mathbf{u}^*\{\mathbf{M}(\lambda_1) - \mathbf{M}(\lambda_2)\}\mathbf{u} \geqslant 0$ for $\lambda_1 \geqslant \lambda_2$ and $\mathbf{M}(\lambda)$ is a matrix having $m_{ij}(\lambda)$ in the ith row and jth column. For example, taking $u_i = 1$; $u_j = 0$, $j \neq i$ we have

$$\rho_{ii}(t) = \int_{-\pi}^{\pi} e^{it\lambda}\, dm_{ii}(\lambda)$$

where $m_{ii}(\lambda)$ is $G_u(\lambda)$ for this particular vector $\mathbf{u}$. Thus we may write

$$\mathbf{R}(h) = \int_{-\pi}^{\pi} e^{ih\lambda}\, d\mathbf{M}(\lambda) \tag{1}$$

where the right-hand side is a shorthand notation for the matrix

$$\left[\int_{-\pi}^{\pi} e^{ih\lambda}\, dm_{ij}(\lambda)\right].$$

[1] The star here indicates transposition combined with conjugation.

Since $\rho_{ij}(t) = \rho_{ji}(-t)$ it is evident that

$$dm_{ij}(\lambda) = -\,dm_{ji}(-\lambda) = \overline{dm_{ji}(\lambda)}$$

and $d\mathbf{M}(\lambda)$ is Hermitian. We have seen that it is non-negative definite. Grenander calls $\mathbf{M}(\lambda)$ the (matrix) spectral distribution function of the regressor set.

If $\{x_t\}$ is a finite moving average of order q we may evaluate the expression (V.2.2) as follows.

To prevent (V.2.2) from going to zero we introduce the diagonal, $(k \times k)$, matrix $\mathbf{D}_n$ having $Y_{n,j}$ in the jth place in the main diagonal and form

$$\lim_n \mathbf{D}_n^{-1}\mathbf{Y}'\mathbf{\Gamma}_n\mathbf{Y}\mathbf{D}_n^{-1} = \lim_n \left[\sum_{u,v=1}^{n} \frac{y_{i,u}y_{j,v}\gamma_{u-v}}{Y_{n,i}\,Y_{n,j}}\right],$$

where the element in row i, column j is shown,

$$= \lim_n \left[\sum_0^q \gamma_t \frac{\sum_{v=1}^{n-t} y_{i,v+t}y_{j,v}}{Y_{n,i}\,Y_{n,j}} + \sum_{-q}^{-1} \gamma_t \frac{\sum_{u=-t+1}^{n} y_{j,u+t}y_{i,u}}{Y_{n,i}\,Y_{n,j}}\right]$$

$$= \left[\sum_{-q}^{q} \gamma_t \rho_{ij}(t)\right] = \left[\int_{-\pi}^{\pi} \sum_{-q}^{q} \gamma_t e^{it\lambda}\, dm_{ij}(\lambda)\right]$$

$$= \left[2\pi \int_{-\pi}^{\pi} f(\lambda)\, dm_{ij}(\lambda)\right] = 2\pi \int_{-\pi}^{\pi} f(\lambda)\, d\mathbf{M}(\lambda).$$

Since $\lim_n \mathbf{D}_n^{-1}[\mathbf{Y}'\mathbf{Y}]\mathbf{D}_n^{-1} = \mathbf{R}(0)$ we have

$$\lim_n \mathbf{D}_n \mathscr{E}\{(\hat{\boldsymbol{\delta}} - \boldsymbol{\delta})(\hat{\boldsymbol{\delta}} - \boldsymbol{\delta})'\}\mathbf{D}_n = \mathbf{R}(0)^{-1} 2\pi \int_{-\pi}^{\pi} f(\lambda)\, d\mathbf{M}(\lambda)\mathbf{R}(0)^{-1}. \quad (2)$$

If x_t has an absolutely continuous spectral function with continuous $f(\lambda)$ then we may find two real trigonometric polynomials $f_1(\lambda)$ and $f_2(\lambda)$ for which

$$f_1(\lambda) \leqslant f(\lambda) \leqslant f_2(\lambda)$$

and

$$\{f_2(\lambda) - f_1(\lambda)\} < \varepsilon.$$

Then if $\Gamma_n^{(1)}$, Γ_n and $\Gamma_n^{(2)}$ are the three corresponding covariance matrices we have, for any vector *u*,

$$\mathbf{u}^*[\Gamma_n - \Gamma_n^{(1)}]\mathbf{u} = \int_{-\pi}^{\pi} \left| \sum_1^n u_j e^{ij\lambda} \right|^2 \{f(\lambda) - f_1(\lambda)\}\, d\lambda \geqslant 0$$

$$\mathbf{u}^*[\Gamma_n - \Gamma_n^{(2)}]\mathbf{u} = \int_{-\pi}^{\pi} \left| \sum_1^n u_j e^{ij\lambda} \right|^2 \{f(\lambda) - f_2(\lambda)\}\, d\lambda \leqslant 0.$$

Thus in particular

$$\mathbf{u}^*\mathbf{D}_n[\mathbf{Y}'\mathbf{Y}]^{-1}\mathbf{D}_n\mathbf{D}_n^{-1}\mathbf{Y}'\Gamma_n\mathbf{Y}\mathbf{D}_n^{-1}\mathbf{D}_n[\mathbf{Y}'\mathbf{Y}]^{-1}\mathbf{D}_n\mathbf{u}$$
$$\geqslant \mathbf{u}^*\mathbf{D}_n[\mathbf{Y}'\mathbf{Y}]^{-1}\mathbf{D}_n\mathbf{D}_n^{-1}\mathbf{Y}'\Gamma_n^{(1)}\mathbf{Y}\mathbf{D}_n^{-1}\mathbf{D}_n[\mathbf{Y}'\mathbf{Y}]^{-1}\mathbf{D}_n\mathbf{u}$$

and therefore, by (2) above,

$$\liminf_n \mathbf{u}^*\mathbf{D}_n\mathscr{E}\{(\hat{\boldsymbol{\delta}} - \boldsymbol{\delta})(\hat{\boldsymbol{\delta}} - \boldsymbol{\delta})'\}\mathbf{D}_n\mathbf{u}$$
$$\geqslant \mathbf{u}^*\mathbf{R}(0)^{-1}\int_{-\pi}^{\pi} f_1(\lambda)\, d\mathbf{M}(\lambda)\mathbf{R}(0)^{-1}\mathbf{u}.$$

Similarly

$$\limsup_n \mathbf{u}^*\mathbf{D}_n\{\mathscr{E}(\hat{\boldsymbol{\delta}} - \boldsymbol{\delta})(\hat{\boldsymbol{\delta}} - \boldsymbol{\delta})'\}\mathbf{D}_n\mathbf{u}$$
$$\leqslant \mathbf{u}^*\mathbf{R}(0)^{-1}\int_{-\pi}^{\pi} f_2(\lambda)\, d\mathbf{M}(\lambda)\mathbf{R}(0)^{-1}\mathbf{u}.$$

Since the difference between the two right-hand sides may be made arbitrarily small by decreasing ε we see that (2) holds in this case also. The conditions on $f(\lambda)$ may be considerably relaxed no doubt, but will suffice for our purposes.

In an analogous manner, if $f(\lambda) \neq 0$ a.e., we may approximate to $f(\lambda)^{-1}$ above and below by spectral densities of processes of autoregressive type and obtain for the expression (V.2.3) the limiting form,

$$\lim_n \mathbf{D}_n\mathscr{E}\{(\tilde{\boldsymbol{\delta}} - \boldsymbol{\delta})(\tilde{\boldsymbol{\delta}} - \boldsymbol{\delta})'\}\mathbf{D}_n = \left\{\frac{1}{2\pi}\int_{-\pi}^{\pi} f(\lambda)^{-1}\, d\mathbf{M}(\lambda)\right\}^{-1}. \quad (3)$$

It is convenient to put $\mathbf{N}(\lambda) = \mathbf{R}(0)^{-\frac{1}{2}}\{\mathbf{M}(\lambda+) - \mathbf{M}(-\lambda-)\}\mathbf{R}(0)^{-\frac{1}{2}}, \lambda \geqslant 0.$

(Since $\mathbf{R}(0)$ is positive definite the 'positive' square root may be uniquely defined – see Halmos [2] p. 136.) We thus consider

$$\lim_n [\mathbf{D}_n \mathscr{E}\{(\hat{\boldsymbol{\delta}} - \boldsymbol{\delta})(\hat{\boldsymbol{\delta}} - \boldsymbol{\delta})'\}\mathbf{D}_n][\mathbf{D}_n \mathscr{E}\{(\tilde{\boldsymbol{\delta}} - \boldsymbol{\delta})(\tilde{\boldsymbol{\delta}} - \boldsymbol{\delta})'\}\mathbf{D}_n]^{-1}$$

$$= \mathbf{R}(0)^{-\frac{1}{2}} \int_{0-}^{\pi} f(\lambda)\, d\mathbf{N}(\lambda) \int_{0-}^{\pi} f(\lambda)^{-1}\, d\mathbf{N}(\lambda)\mathbf{R}(0)^{\frac{1}{2}}. \quad (4)$$

Now as we shall presently see there are situations (i.e. regressor sets $\mathbf{y}_j$) where $\mathbf{N}(\lambda)$ increases only at $q \leqslant k$ points λ_j, the jth increase being $\mathbf{N}_j$, let us say, and these $\mathbf{N}_j$ satisfy

$$\mathbf{N}_i\mathbf{N}_j = \mathbf{O} \text{ (the null matrix)} \quad i \neq j.$$

Since $\sum \mathbf{N}_j = \mathbf{I}_k$ it follows that $\mathbf{N}_j^2 = \mathbf{N}_j$ (and the $\mathbf{N}_j$, being Hermitian, are therefore orthogonal projections).

In this case (4) becomes

$$\mathbf{R}(0)^{-\frac{1}{2}} \sum_j f(\lambda_j)\mathbf{N}_j \sum_j f(\lambda_j)^{-1}\mathbf{N}_j\mathbf{R}(0)^{\frac{1}{2}}$$

$$= \mathbf{R}(0)^{-\frac{1}{2}} \sum_j f(\lambda_j)f(\lambda_j)^{-1}\mathbf{N}_j\mathbf{R}(0)^{\frac{1}{2}} = \mathbf{I}_k$$

and thus the least squares and b.l.u. estimates have, asymptotically, the same covariance matrices.

The points λ_j are called by Grenander the elements of the spectrum of the regressor set.[1]

There are two cases of prime importance where this condition on the $\mathbf{N}(\lambda)$ is satisfied.

(i) Consider the case where the k regressor vectors correspond to a polynomial trend of degree $k - 1$. Then it is not difficult to see that the correlation of t^r and $(t + h)^s$ converges to

$$\sqrt{\{(2r + 1)(2s + 1)\}}/(r + s + 1)$$

and is thus independent of h. Thus $\mathbf{M}(\lambda)$ has only one point of increase, at the origin, and q in the theorem given above is unity, the

[1] In general the interval $[0, \pi]$ can be decomposed into Lebesgue measurable sets s_j such that the increments of $\mathbf{N}(\lambda)$ over these sets satisfy the orthogonality condition. The sets s_j are called the elements of the spectrum. See Grenander and Rosenblatt [2].

single element of the spectrum of the regressor set being a point. Least squares is thus asymptotically fully efficient. It is to be noted that the theorem was proved upon the assumption that $f(\lambda)$ was not zero. No doubt this condition could be relaxed in some circumstances, but there is one situation where it certainly cannot and this is the case where the zero coincides with a point where a finite spectral mass of the regressor set is concentrated. Just this effect will be created if we difference all of the variables when a polynomial trend is being removed, in case x_t is generated by a stationary process, for the differencing will then introduce a zero at the origin into the spectral density of the differenced residuals. In this situation, as we have already seen in the example 2 of section 2, the least squares procedure will have zero asymptotic efficiency.

It is easily checked that the covariance matrix of the least squares estimates of the δ_j will be of the form

$$2\pi f(0)[n^{r+s+1}/(r+s+1)]^{-1}$$

where the element corresponding to the regression coefficients of t^r and t^s is shown.

If we use orthogonal polynomials so that the regressor set is orthonormalized then for large n the covariance matrix will be

$$2\pi f(0)\mathbf{I}_k.$$

(ii) The simplest case of trigonometric regression involves a mean value for z_t of the form ($k-1$ even)

$$\alpha_0 + \sum_{1}^{\frac{1}{2}(k-1)} \alpha_j \cos \lambda_j t + \sum_{1}^{\frac{1}{2}(k-1)} \beta_j \sin \lambda_j t.$$

Here the matrix valued function $\mathbf{N}(\lambda)$ will have $\frac{1}{2}(k+1)$ points of increase; at the origin and the points λ_j. The covariance matrix will be diagonal with $(4\pi/n)f(\lambda_j)$ corresponding to α_j, or β_j, in the main diagonal, and $(2\pi/n)f(0)$ corresponding to α_0.

Thus for polynomial or trigonometric regression, for large n at least, the straightforward least squares procedure is efficient under quite general conditions. This result, due to Grenander, is certainly a pleasant one. Of course the covariance matrix of the $\hat{\boldsymbol{\delta}}$ still depends upon $f(\lambda)$ so that we may still wish to test the significance of the serial correlations in the residuals. This test may also, however, be simplified for a polynomial regression, as we shall see.

The formulae (2) and (3) are of general interest for they indicate precisely how the efficiency of the least squares procedure may differ from unity and why it does so. If $f(\lambda)$ is not far from uniform then for '*most*' $\mathbf{M}(\lambda)$ the loss of efficiency from the use of least squares should not be great and this should, in particular, be so if $\mathbf{M}(\lambda)$ is absolutely continuous, as would be the case if the regressor set was obtained from a strictly stationary process with an absolutely continuous spectrum. Thus in such conditions the use of a transformation which will make $f(\lambda)$ *nearly* uniform may be a worthwhile device.

(*b*) Testing the independence of the residuals.

We have seen, in the previous section, that in the case (ii) above the bounds test for serial correlation of the residuals may be replaced by an exact test, at least in the cases covered by the tabulations in Anderson and Anderson [1]. When the regressor vectors are those corresponding to a polynomial trend the bounds test may again be improved upon. The considerations of the last section show that either bound is an approximation, in any case, to the true significance point, if we neglect quantities of order n^{-1}. However, it may be shown (Hannan [2]) that the *upper bound in the case of a polynomial regression*, is an approximation to the true significance if we neglect only quantities of order n^{-2}. The approximation is in fact extremely good even for n small and k large. For example for $n = 15$ and $k = 6$ (a 5th degree polynomial) the true significance point at the 5% level (for a one-sided test) will be 2·15 while the upper bound is 2·21. (The lower bound in this case is 0·56 so that the bounds are too wide to be of much use.) For $n = 30$ and $k = 6$ the approximation will involve an error of about 0·01.

The derivation of this approximation follows from a consideration of the matrix

$$\mathbf{Q}'\mathbf{B}_d\mathbf{Q}, \quad \mathbf{Q} = \mathbf{I}_n - \mathbf{Y}[\mathbf{Y}'\mathbf{Y}]^{-1}\mathbf{Y}'.$$

The elements of $\mathbf{B}_d$ are of the form

$$b_{jk} = \frac{1}{2\pi}\int_{-}^{\pi} g(\lambda)e^{i(j-k)\lambda}\,d\lambda$$

where $g(\lambda) = 2(1 - \cos\lambda)$, (except for b_{11} and b_{nn}), and it follows (cf. section I.5) that the proper values of $\mathbf{B}_d$, $\lambda_{n,j} = 2(1 - \cos\frac{1}{2}\psi_j)$, $= 0, \ldots, n-1$; satisfy

$$\lim_{n\to\infty} n^{-1} \operatorname{Tr}(\mathbf{B}_d^s) = \lim_{n\to\infty} n^{-1} \sum \lambda_{n,j}^s = \frac{1}{2\pi}\int_{-\pi}^{\pi} \{g(\lambda)\}^s \, d\lambda.$$

It may be shown (Hannan [2]) that

$$\lim_{n\to\infty} n^{-1} \operatorname{Tr}[\mathbf{Q}'\mathbf{B}_\alpha\mathbf{Q}]^s = \lim_{n\to\infty} n^{-1} \operatorname{Tr}[\mathbf{Q}\mathbf{B}_\alpha^s]$$

and making use of the fact that the trace of a product of two factors is independent of their order, this is

$$\lim_{n\to\infty} n^{-1}\{\operatorname{Tr}(\mathbf{B}_\alpha^s) - \operatorname{Tr}(\mathbf{Y}'\mathbf{B}_\alpha^s\mathbf{Y}[\mathbf{Y}'\mathbf{Y}]^{-1})\}$$

However,

$$\mathbf{D}_n^{-1}\mathbf{Y}'\mathbf{B}_d^s\mathbf{Y}\mathbf{D}_n^{-1} \to \int_{-\pi}^{\pi} \{g(\lambda)\}^s \, d\mathbf{M}(\lambda)$$

by the same argument which led to (2). Thus

$$\operatorname{Tr}\{\mathbf{Y}'\mathbf{B}_d^s\mathbf{Y}[\mathbf{Y}'\mathbf{Y}]^{-1}\} = \operatorname{Tr}\{\mathbf{D}_n(\mathbf{D}_n^{-1}\mathbf{Y}'\mathbf{B}_d^s\mathbf{Y}\mathbf{D}_n^{-1})\mathbf{D}_n[\mathbf{Y}'\mathbf{Y}]^{-1}\}$$
$$= \operatorname{Tr}\{\mathbf{D}_n[\mathbf{Y}'\mathbf{Y}]^{-1}\mathbf{D}_n\mathbf{D}_n^{-1}\mathbf{Y}'\mathbf{B}_d^s\mathbf{Y}\mathbf{D}_n^{-1}\} \to$$
$$\operatorname{Tr}\left\{\mathbf{R}(0)^{-1}2\pi\int_{-\pi}^{\pi} g(\lambda)^s \, d\mathbf{M}(\lambda)\right\}.$$

Thus

$$n^{-1}\operatorname{Tr}[\mathbf{Q}\mathbf{B}_d\mathbf{Q}]^s \to \frac{1}{2\pi}\left[\int_{-\pi}^{\pi} \{g(\lambda)\}^s \, d\lambda - \frac{2\pi}{n}\operatorname{Tr}\left\{\mathbf{R}(0)^{-\frac{1}{2}}\int_{-\pi}^{\pi} g(\lambda)^s \, d\mathbf{M}(\lambda)\mathbf{R}(0)^{-\frac{1}{2}}\right\}\right].$$

If the elements of the spectrum of the regressor vectors are points λ_j this becomes

$$\frac{1}{2\pi}\left\{\int_{-\pi}^{\pi} g(\lambda)^s \, d\lambda - \frac{2\pi}{n}\operatorname{Tr}\left(\sum_j g(\lambda_j)^s\mathbf{N}_j\right)\right\}$$
$$= \frac{1}{2\pi}\left\{\int_{-\pi}^{\pi} g(\lambda)^s \, d\lambda - \frac{2\pi}{n}\sum p_j g(\lambda_j)^s\right\}$$

where p_j (an integer) is Tr $\mathbf{N}_j$ and $\sum p_j = k$. Since the moments of the test statistic may be expressed solely in terms of the quantities Tr $[\mathbf{QB}_d\mathbf{Q}]^s$ [1] it follows that the distribution of this statistic will be, to order n^{-1}, the same as that which would obtain if the regressor set had been composed of proper vectors of $\mathbf{B}_d$ corresponding to the proper values $g(\lambda_j)$ (repeated p_j times). In particular for a regression on a polynomial the effect is to remove from the spectrum of $\mathbf{B}_d$ the smallest proper values so that the Durbin and Watson upper bound to the significance point for the test statistic becomes appropriate.

There is one further case where this theory becomes appropriate. This occurs when a mixed regression is being calculated involving, let us say k_1 regressor vectors of general type and $k - k_1 = k_2$ additional regressors, the elements of whose spectrum consist of $q \leqslant k_2$ points. The situation we have in mind is that where a polynomial trend is being removed, from all variables involved, by the inclusion of polynomial terms in the regression. In this case k can be quite high and can result in bounds for the test statistic being uncomfortably wide. We orthonormalize the k_2 vectors of the second set, indicating the orthonormalized vectors by $\boldsymbol{\varphi}_j$, and then adjoin k_1 further orthonormal vectors $\mathbf{w}_j$ so that the whole set spans the same space as that spanned by the original $\mathbf{y}_j$. Then $\mathbf{QB}_d\mathbf{Q}$ becomes $\mathbf{Q}_1\mathbf{Q}_2\mathbf{B}_d\mathbf{Q}_2\mathbf{Q}_1$ where $\mathbf{Q}_2$ corresponds to the $\boldsymbol{\varphi}_j$ and $\mathbf{Q}_1$ to the $\mathbf{w}_j$. If the proper values of $\mathbf{Q}_2\mathbf{B}_d\mathbf{Q}_2$ are $\mu_0 \leqslant \mu_1 \leqslant \ldots \leqslant \mu_{n-k_2-1}$ while those of $\mathbf{QB}_d\mathbf{Q}$ are, $\nu_0 \leqslant \nu_1 \leqslant \ldots \leqslant \nu_{n-k-1}$, then

$$\mu_i \leqslant \nu_i \leqslant \mu_{i+k_1}.$$

Thus a lower bound to r_d is

$$\sum_0^{n-k-1} \mu_i \xi_i^2 \Big/ \sum_0^{n-k-1} \xi_i^2. \tag{5}$$

However,

$$\lim_n n^{-1} \sum_0^{n-k-1} \mu_i^s$$

[1] See section V.3.

$$= \lim_n n^{-1}\left\{\sum_0^{n-k_2-1} \mu_i^s - \sum_1^{k_1} \mu_{n-k_2-j}^s\right\}$$

$$= \frac{1}{2\pi}\left[\int_{-\pi}^{\pi} g(\lambda)^s\, d\lambda - \frac{2\pi}{n}\sum_1^{k_2} p_j g(\lambda_j)^s - \frac{2\pi}{n}\sum_1^{k_1} \mu_{n-k_2-j}^s\right]$$

with
$$\sum p_j = k_2.$$

Since $\sum_1^{k_1} \mu_{n-k_2-j}^s$ will differ from $\sum_1^{k_1} \lambda_{n,\,n-j}^s$ by a quantity of order n^{-1} [1] this is, to the order n^{-1}

$$\frac{1}{2\pi}\int_{-\pi}^{\pi} g(\lambda)^s\, d\lambda - \frac{2\pi}{n}\sum_1^{k_2} p_j g(\lambda_j)^s - \frac{2\pi}{n}\sum_1^{k_1} \lambda_{n,\,n-j}^s.$$

Thus (5) is, to the order n^{-1}, distributed as if the regression had been upon a set of vectors of which k_1 corresponded to the greatest proper values of $\mathbf{B}_d$ while the remaining k_2 were proper vectors corresponding to the proper values $g(\lambda_j)$ (repeated p_j times). In the case of a polynomial set providing the k_2 vectors, the bound to the significance point becomes that appropriate to

$$\sum_{k_2-1}^{n-k_1-1} \lambda_{n,\,i}\xi_i^2 \Big/ \sum_{k_2-1}^{n-k_1-1} \xi_i^2 \tag{6}$$

since the k_2 $g(\lambda_j)$ which occur are then, to order n^{-1}, the k_2 smallest $\lambda_{n,j}$.

This statistic is not tabulated but if we add back to numerator and denominator the terms corresponding to the smallest roots (other than the term corresponding to $\lambda_{n,0} = 0$) we obtain a statistic of the form

$$\sum_1^{n-k_1-1} \lambda_{n,\,i}\xi_i^2 \Big/ \sum_1^{n-k_1-1} \xi_i^2$$

[1] Since $\lambda_{n,i} \leq \mu_i \leq \lambda_{n,\,i+k_2}$ and $\lambda_{n,\,i}$ and $\lambda_{n,\,i+k_2}$ differ by $O(n^{-1})$.

whose significance points will provide a lower bound to (6)[1] and thus, finally, to r_d. A comparison with the formula (3) of section (3) shows that in a mixed regression we may use as bounds to the significance points for r_d the upper bound as tabulated by Durbin and Watson and the lower bound obtained by entering their tables as if the regressor set had consisted of $k_1 + 1$ vectors only, one of which corresponds to the mean. This will usually materially reduce the width of the bounds. For example with $n = 37$, $k_1 = 1$ and $k_2 = 5$ (a 4th degree polynomial) the bounds using $k = 6$ in Durbin and Watson's tables are 1·19 and 1·80. Entering the tables with $k = 2$ (then $k' = k - 1 = 1$) we find that a lower bound is 1·42, which materially improves upon 1·19. An added advantage of this modification lies in the fact that it enables us to locate an (approximate) lower bound from the Durbin and Watson tables in cases where these tables would not otherwise be sufficiently extensive ($k = 6$ is the largest value tabulated).

Example 1

The following 15 observations, z_t, are generated by a simple Markoff process $z_t - 0{\cdot}5z_{t-1} = \varepsilon_t$ where the ε_t are standard normal with zero mean.

t	z_t	t	z_t	t	z_t	t	z_t
1	−0·30.	5	1·20	9	0·15.	13	0·62
2	−1·12.	6	1·70.	10	−0·66.	14	0·12
3	1·82.	7	0·82.	11	−1·50.	15	−0·83
4	1·69.	8	−1·06.	12	−0·80.		

If we fit a trend $\delta_1 + \delta_2 t$ we obtain the least squares estimate $\delta_1 = 0{\cdot}86$, $\delta_2 = -\ 0{\cdot}09$ while the first differences give the estimate $\delta_2{}^* = -\ 0{\cdot}04$. The asymptotic formulae for the standard deviations of δ_2 and $\tilde{\delta}_2$ give the estimates of the respective standard deviations as 0·06 and 0·11 so that the fact that the less efficient estimate gives a more precise figure for δ_2 in this case is in no way surprising. The statistic r_d is 1·37 which lies almost upon the upper bound to the 5%

[1] This lower bound will *not* be a strict lower bound because of the approximations involved but the error will be too small to matter in practice.

point (1·36). In this case the lower bound is 1·08 which is a fair way from the true significance point (which *is* 1·36 in this case).

Example 2

Quenouille [3] p. 183 investigates the relation between U.S. fertilizer consumption z_t and gross farm income y_t after the removal of a trend, described by a polynomial of degree 5, from both series. The first serial correlation of the residuals from the regression of z_t on the remaining 7 variables is 0·229 corresponding to a value of $r_d = 1{\cdot}54$. Here $n = 37$, $k - 1 = 6$ so that no bounds are available from the Durbin and Watson tables. The Durbin and Watson bounds for a one-sided test at the 5% level are in fact 1·14 and 1·88. The lower bound from (6) is 1·42. Thus even with the improved lower bound no firm conclusion is reached since the observed value falls between the bounds. For this case the untabulated statistic (5) has a 5% point equalling 1·75 so that there is no doubt that the result is significant, as Quenouille conjectured. Quenouille proceeded by carrying out a new regression in which he included the terms z_{t-1} and y_{t-1}. The serial correlation of the resulting residuals was then 0·063 corresponding to $r_d = 1{\cdot}87$. The results of this section are, however, now not applicable since z_{t-1} is certainly not independent of the residuals x_t.

5. The effect of trend removal on the analysis of the residuals

In Chapters II, III and IV our analysis was based upon the assumption that the mean value of the process being examined was zero, though in some cases where it was convenient, mean corrections were made. The derivations of the results of these sections were, in general, based upon the assumption that x_t was generated by a linear process with finite κ_4 for which, at least, $\sum |\alpha_j| < \infty$. Under these conditions it may be shown by straightforward methods that

$$(n - t)b_n(t) = \sum_1^{n-t} \hat{x}_s\hat{x}_{s+t} - \sum_1^{n-t} x_s x_{s+t}$$

is $O(1)$ uniformly in t (in the sense that its mean square is $O(1)$).

Thus the bias in c_t due to regression is $O((n-t)^{-1})$. This will in turn ensure that the effects of the regression will be asymptotically small compared with terms retained in formulas derived in earlier chapters, when no regression was carried out. For example the bias due to regression in the estimate of $f(\lambda)$ obtained by the methods of section III.2 will be

$$\frac{1}{2\pi}\sum_{-m_n}^{m_n}\left(1-\frac{|t|}{n}\right)b_n(t)k(t/m_n)e^{it\lambda}$$

which under the conditions of that section will be $O(m_n/n)$. If $\sum|\gamma_t t^q|<\infty$ and m_n is chosen, optimally, to be of order $n^{1/(1+2q)}$ then the mean square error of the estimate of $f(\lambda)$ is $O(n^{-2q/(1+2q)})$ while the mean square of the bias due to regression will be only $O(n^{-4q/(1+2q)})$.

Nevertheless the effects of the regression may be large in small samples. Indeed we have (for orthonormalized $\mathbf{y}_j$)

$$\tfrac{1}{2}\mathscr{E}\{I_n(\lambda,\hat{x})\}=\sum_{-n+1}^{n-1}\gamma_t\left(1-\frac{|t|}{n}\right)e^{it\lambda}-n^{-1}\mathbf{l}^*\mathbf{\Gamma}_n\mathbf{Y}\mathbf{Y}'\mathbf{l}$$
$$-n^{-1}\mathbf{l}^*\mathbf{Y}\mathbf{Y}'\mathbf{\Gamma}_n\mathbf{l}+n^{-1}\mathbf{l}^*\mathbf{Y}\mathbf{Y}'\mathbf{\Gamma}_n\mathbf{Y}\mathbf{Y}'\mathbf{l} \qquad (1)$$

where the vector $\mathbf{l}$ has $e^{it\lambda}$ in the tth place.

Introducing $I_n(\lambda,j)=\left|\sum_t y_{j,t}e^{it\lambda}\right|^2$, the second term in (1) becomes

$$n^{-1}\sum_{j=1}^{k}\left\{I_n(\lambda,j)\int_{-\pi}^{\pi}f(\theta)\sum_1^n e^{it(\lambda-\theta)}\frac{\sum_1^n y_{j,t}e^{it\theta}}{\sum_1^n y_{j,t}e^{it\lambda}}\,d\theta\right\}.$$

This converges to $n^{-1}2\pi f(\lambda)\sum_j I_n(\lambda,j)$. Here the neglected part need not be of higher order than that retained except at points where

$n^{-1}\sum_j I_n(\lambda, j)$ is $O(1)$. However, these are the only points of importance in what follows. The third term in (1) gives a similar contribution while the last is, in the limit

$$n^{-1}\mathbf{l}^*\mathbf{Y}2\pi\int_{-\pi}^{\pi} f(\lambda)\, d\mathbf{N}(\lambda)\mathbf{Y}'\mathbf{l}.$$

Thus the bias introduced by the regression is, asymptotically,

$$-n^{-1}4\pi f(\lambda)\sum_j I_n(\lambda, j) + n^{-1}\mathbf{l}^*\mathbf{Y}2\pi\int_{-\pi}^{\pi} f(\lambda)\, d\mathbf{N}(\lambda)\mathbf{Y}'\mathbf{l}. \qquad (2)$$

If the regressor vectors are obtained by orthonormalizing a set of vectors generated by stationary processes then for any given λ, $\mathbf{l}^*\mathbf{y}_j$ will be $O(1)$, as we have seen in the third chapter, and the expression (2) will be $O(n^{-1})$, so that the bias is everywhere negligible. Interest therefore concentrates upon the case where $\mathbf{N}(\lambda)$ has q points of increase λ_p while $n^{-\frac{1}{2}}\mathbf{l}^*\mathbf{Y}$ converges to zero save at these q points.[1] Then (2) becomes[2]

$$-n^{-1}2\pi f(\lambda)\sum_{j=1}^{k} I_n(\lambda, j)$$

since the last term may be replaced by

$$n^{-1}2\pi\sum_{p=1}^{q} f(\lambda_p)\mathbf{l}^*\mathbf{Y}\mathbf{N}_p\mathbf{Y}'\mathbf{l} \rightarrow n^{-1}2\pi f(\lambda)\sum_{1}^{q}\mathbf{l}^*\mathbf{Y}\mathbf{N}_p\mathbf{Y}'\mathbf{l}$$

(the $\mathbf{N}_p$ being the increments in $\mathbf{N}(\lambda)$ at the points λ_p).

If we now consider an estimate of $f(\lambda)$ of the form

$$f_n(\lambda) = \int_{-\pi}^{\pi} w_n(\theta - \lambda) I_n(\theta, \hat{x})\, d\theta$$

[1] Which will, of course, be so for polynomial or trigonometric regression.

[2] This formula may be used to obtain an approximation, to order n^{-1}, to the bias in the c_j and r_j. See Hannan [3].

then

$$\mathscr{E}\{\hat{f}_n(\lambda)\} \to 4\pi \int_{-\pi}^{\pi} w_n(\theta - \lambda) f(\theta) \left\{ 1 - n^{-1} \sum_j I_n(\theta, j) \right\} d\theta$$

$$\approx 4\pi \int_{-\pi}^{\pi} w_n(\theta - \lambda) f(\theta)\, d\theta - 4\pi n^{-1} f(\lambda) \sum \int_{-\pi}^{\pi} w_n(\theta - \lambda) I_n(\theta, j)\, d\theta$$

if $f(\lambda)$ is a reasonably smooth function of λ.[1] The first term here will be $f(\lambda)$ except for a bias term arising from the smoothing operation on $I_n(\lambda, \hat{x})$. The second term will be small save in the immediate neighbourhood of the points where the jumps in $N(\lambda)$ occur. As we have already seen, for an optimal choice of m_n relative to the rate of convergence of the γ_t, this bias term cannot matter asymptotically even at the points λ_p. However, in small samples it is likely to be appreciable here.[2] Indeed its size can be gauged from the size of the known factor $4\pi n^{-1} \sum_j \int_{-\pi}^{\pi} w_n(\theta - \lambda) I_n(\theta, j)\, d\theta$. If this is large then a worthwhile modification of $\hat{f}_n(\lambda)$ may be obtained by forming

$$\left\{ 1 - 4\pi n^{-1} \sum_j \int_{-\pi}^{\pi} I_n(\theta, j) w_n(\theta - \lambda)\, d\theta \right\}^{-1} \hat{f}_n(\lambda) = h(\lambda)^{-1} \hat{f}_n(\lambda), \quad (3)$$

let us say.

The first factor here is totally analogous to the factor $(1 - n^{-1})^{-1}$ by which the estimate $n^{-1} \sum_1^n (x_t - \bar{x})^2$ is multiplied, in order to obtain an unbiased estimate of the variance of the x_t when these variables are independent. The importance of this factor of course depends upon k and while it is impossible to prescribe any precise mode of variation of k with n it is probable that for large n, k will be relatively large also.

[1] If it were not there there would be little point in attempting to estimate it by a smoothing process.

[2] As has already been seen, these points are important as the value of $f(\lambda)$ at them enters as a factor in the variance of the estimate of the regression coefficients.

If two of the regressor vectors correspond to $\cos t\psi_j$ and $\sin t\psi_j$ then the two corresponding terms in $\sum_j I_n(\lambda, j)$ sum to

$$\left\{\frac{\sin^2 \frac{1}{2}n(\lambda + \psi_j)}{n \sin^2 \frac{1}{2}(\lambda + \psi_j)} + \frac{\sin^2 \frac{1}{2}n(\lambda - \psi_j)}{n \sin^2 \frac{1}{2}(\lambda - \psi_j)}\right\}.$$

If m_n is small relative to n then the corresponding component in

$$4\pi n^{-1} \sum \int_{-\pi}^{\pi} I_n(\theta, j) w_n(\theta - \lambda)\, d\theta \tag{4}$$

is, approximately,

$$8\pi^2 n^{-1}\{w_n(\lambda + \psi_j) + w_n(\lambda - \psi_j)\}. \tag{5}$$

For a set of q regressor vectors generated by a polynomial of degree $q - 1$ the corresponding contribution to (4) is approximately,

$$8\pi^2(q/n) w_n(\lambda). \tag{6}$$

Thus in all of the cases treated in section III.2 the expression (4) may be approximated, in the circumstances which arise in practice, easily and satisfactorily so that one can consider whether the first factor in (3) should be applied.

Example

The data consisted of 60 observations from a simple Markoff process with parameter 0·5. The data is given in Jowett [1] p. 218. A trend of the form

$$\delta_0 + \delta_1 t + \delta_2 \cos t\frac{\pi}{2} + \delta_3 \sin t\frac{\pi}{2}$$

was fitted and the spectral density estimated using the truncated formula with $m_n = 5$. The factor $h(\lambda)$ is then, by our approximations (5) and (6), since $8\pi^2 w_n(\lambda) = (\sin \frac{1}{2}\lambda)^{-1}\{\sin \frac{1}{2}(2m_n + 1)\lambda\}$,

$$h(\lambda) = \left[1 - \frac{1}{60}\left\{2\frac{\sin \frac{11}{2}\lambda}{\sin \frac{1}{2}\lambda} + \left(\frac{\sin \frac{11}{2}\left(\lambda + \frac{\pi}{2}\right)}{\sin \frac{1}{2}\left(\lambda + \frac{\pi}{2}\right)} + \frac{\sin \frac{11}{2}\left(\lambda - \frac{\pi}{2}\right)}{\sin \frac{1}{2}\left(\lambda - \frac{\pi}{2}\right)}\right)\right\}\right].$$

The results are shown below for the interval $[0, \pi]$ in Table 7.

TABLE 7

λ, degrees	$f(\lambda)$	$\hat{f}_n(\lambda)$	$h(\lambda)$	$h(\lambda)^{-1}\hat{f}_n(\lambda)$
0	·48	·31	·60	·50
10	·45	·31	·66	·47
20	·39	·32	·83	·39
30	·31	·31	1·00	·31
40	·25	·28	1·10	·23
50	·20	·22	1·10	·20
60	·16	·15	1·00	·15
70	·13	·10	·88	·11
80	·11	·08	·80	·10
90	·10	·09	·80	·11
100	·08	·12	·86	·14
110	·08	·13	·94	·14
120	·07	·12	1·00	·12
130	·06	·10	1·02	·10
140	·06	·07	1·03	·07
150	·05	·07	1·00	·07
160	·05	·08	1·00	·08
170	·05	·10	·99	·10
180	·05	·11	1·00	·11

The improvement in the estimate due to the application of $h(\lambda)$ is obvious. It is to some extent hidden in the neighbourhood of 100 degrees and 180 degrees by oscillations due to the nature of the weight function (which has a secondary peak about 80 degrees away from the main one).

This case is favourable to the use of the factor $h(\lambda)$ since the bias in $\hat{f}_n(\lambda)$ due to smoothing is very small. It should not therefore be taken as necessarily indicative of the results to be obtained in general.

Appendix

The Spectral Representation of x_t

If X is a space upon which a measure μ is defined[1] and if $f_n(x)$ is a sequence of functions square integrable with respect to μ then we may ask under what conditions there exists a square integrable function $f(x)$ for which

$$\int_X |f(x) - f_n(x)|^2 \, d\mu \longrightarrow 0.$$

It is a classic[2] fact that the necessary and sufficient condition is

$$\lim_{m, n \to \infty} \int_X |f_m(x) - f_n(x)|^2 \, d\mu = 0.$$

We shall say that such a sequence converges in the mean, with weighting $d\mu$.

To establish the spectral representation we set up a correspondence

$$x_t(\omega) \longleftrightarrow e^{it\lambda} \quad -\pi \leqslant \lambda \leqslant \pi. \tag{1}$$

We then extend this correspondence by linearity so that

$$\sum_p^q \alpha_j x_{t_j}(\omega) \longleftrightarrow \sum_p^q \alpha_j e^{it_j\lambda}. \tag{2}$$

The $\sum_p^q \alpha_j x_{t_j}(\omega)$ are functions square integrable with respect to

[1] Or, at least, 'upon a σ-ring of sets in which μ is defined'. See Halmos [1].
[2] See, for example, Halmos [1] p. 177.

μ on Ω while $\sum_p^q \alpha_j e^{it_j\lambda}$ is square integrable with respect to the measure induced by $dF(\lambda)$ on $[-\pi, \pi]$. Moreover

$$\mathscr{E}\left\{\sum_p^q \alpha_j x_{t_j}(\omega) \overline{\sum_r^s \beta_k x_{t_k}(\omega)}\right\}$$

$$= \int_{-\pi}^{\pi} \sum_p^q \alpha_j e^{it_j\lambda} \overline{\sum_r^s \beta_k e^{it_k\lambda}}\, dF(\lambda) \quad (3)$$

and in particular the integrals of the squared moduli of two corresponding expressions, with respect to their relevant measures, are the same.

We may now extend the correspondence to a wider class, $\mathscr{M}$, of functions on $[-\pi, \pi]$ and a, corresponding, wider class, $\mathscr{M}'$, of functions (random variables) on Ω. These classes consist of those functions which are limits in the mean of sequences of elements of the respective forms (2). Evidently (from (3)) to every sequence of one type which converges in the mean there corresponds a sequence of the other type, and conversely, so that $\mathscr{M}$ and $\mathscr{M}'$ will also be in one–one correspondence.[1] The relation (3) will continue to hold when the correspondence is extended to $\mathscr{M}$ and $\mathscr{M}'$.[2]

Among the extended class of functions of λ are those of the form

$$e_{-\pi}(\lambda) \equiv 0, \quad e_{\pi}(\lambda) \equiv 1$$

$$e_{\theta}(\lambda) = \begin{cases} 1 & -\pi < \lambda \leqslant -\pi + \theta \\ 0 & -\pi + \theta < \lambda \leqslant \pi. \end{cases}$$

Now decompose $[-\pi, \pi]$ by means of points, λ_k,

$$-\pi = \lambda_0 < \lambda_1 < \ldots < \lambda_n = \pi$$

with
$$(\lambda_k - \lambda_{k-1}) \leqslant \frac{1}{t}\left(\frac{\varepsilon}{\gamma_0}\right)^{\frac{1}{2}}.$$

[1] We must of course identify elements (of either space of functions) whose difference has a squared modulus which has zero integral with respect to the relevant measure.

[2] See Riesz and Nagy [1] p. 199, for example.

Then

$$\int_{-\pi}^{\pi} \left| e^{it\lambda} - \sum_k e^{it\lambda_k} \{e_{\lambda_k}(\lambda) - e_{\lambda_{k-1}}(\lambda)\} \right|^2 dF(\lambda) \leqslant \varepsilon$$

since, for $\lambda_{k-1} < \lambda \leqslant \lambda_k$,

$$| e^{it\lambda} - e^{it\lambda_k} |^2 \leqslant t^2 | \lambda - \lambda_k |^2 \leqslant \varepsilon \gamma_0^{-1}.$$

Thus

$$\mathscr{E}\left\{ \left| x_t - \sum_k e^{it\lambda_k} \{z(\lambda_k, \omega) - z(\lambda_{k-1}, \omega)\} \right|^2 \right\} \leqslant \varepsilon$$

where $z(\lambda_k, \omega)$ corresponds to $e_{\lambda_k}(\lambda)$.

Allowing n to increase so that ε tends to zero we are led to put

$$x_t = \int_{-\pi}^{\pi} e^{it\lambda}\, dz(\lambda, \omega)$$

which we have previously written, dropping the ω argument variable for simplicity of notation,

$$x_t = \int_{-\pi}^{\pi} e^{it\lambda}\, dz(\lambda).$$

Since

$$\int_{-\pi}^{\pi} \{e_{\theta_1}(\lambda) - e_{\theta_2}(\lambda)\}\{e_{\theta_3}(\lambda) - e_{\theta_4}(\lambda)\}\, dF(\lambda) = 0$$

for

$$\theta_1 \geqslant \theta_2 > \theta_3 \geqslant \theta_4$$

while

$$\int_{-\pi}^{\pi} \{e_\theta(\lambda)\}^2\, dF(\lambda) = \int_{-\pi}^{\theta} dF(\lambda) = F(\theta)$$

we see from (3) that $z(\lambda)$ is a process of orthogonal increments, having $\mathscr{E}\{| z(\lambda) |^2\} = F(\lambda)$, as required.

In section (I.3) we have formed expressions of the type

$$y(\theta) = \int_{-\pi}^{\theta} g(\lambda)\, dz(\lambda) \tag{4}$$

where $g(\lambda)$ is a polynomial in $e^{i\lambda}$ and $e^{-i\lambda}$ which is never zero. Since the function $g_\theta(\lambda)$, which is $g(\lambda)$ in the range $[-\pi, \theta]$ and zero outside of it (which evidently lies in $\mathscr{M}$), may be approximated uniformly by a function of the type $\sum_k g_\theta(\lambda_k)\{e_{\lambda_k}(\lambda) - e_{\lambda_{k-1}}(\lambda)\}$ it is evident that the proof of the spectral representation formula for x_t carries over directly to give (4) its precise meaning.

Finally since $g(\lambda_k)\{e_{\lambda_k}(\lambda) - e_{\lambda_{k-1}}(\lambda)\}$ may be made uniformly close to $g_{\lambda_k}(\lambda) - g_{\lambda_{k-1}}(\lambda)$ by choosing a fine enough decomposition of $[-\pi, \pi]$, it follows that

$$\sum_k e^{it\lambda_k}\{e_{\lambda_k}(\lambda) - e_{\lambda_{k-1}}(\lambda)\} - \sum_k e^{it\lambda_k} g(\lambda_k)^{-1}\{g_{\lambda_k}(\lambda) - g_{\lambda_{k-1}}(\lambda)\}$$

converges, with n, to zero, in the mean with weighting $dF(\lambda)$. This implies in turn that

$$x_t = \int_{-\pi}^{\pi} e^{it\lambda} g(\lambda)^{-1}\, dy(\lambda)$$

which justifies the substitutions made in section (I.3).

Bibliography

AITKEN, A. C. [1] 'On least squares and the linear combination of observations', *Proc. Roy. Soc. Edinburgh*, **55** (1935) 42–48

ANDERSON, R. L. [1] 'Distribution of the serial correlation coefficient', *Ann. Math. Statist.*, **13** (1942) 1–13

ANDERSON, R. L. and ANDERSON, T. W. [1] 'Distribution of the circular serial correlation for residuals from a fitted Fourier series', *Ann. Math. Statist*, **21** (1950) 59–81

ANDERSON, T. W. [1] 'On the theory of testing serial correlation', *Skand. Aktuarietidskr.*, **31** (1948) 88–116

BARTLETT, M. S. [1] *Stochastic Processes*. Cambridge (1955)

[2] 'On the theoretical specification and sampling properties of autocorrelated time-series', *J.R. Statist. Soc.* (*Suppl.*), **8** (1946) 27–41

[3] 'Problemes de l'analyse spectrale des series temporelles stationnaires', *Publ. Inst. Statist.* (Univ. de Paris), Vol. **3**, Fasc. 3, 119–134

[4] 'Periodogram analysis and continuous spectra', *Biometrika*, **37** (1950) 1–16

[5] 'The problem in statistics of testing several variances', *Proc. Camb. Phil. Soc.*, **30** (1934) 164–169

BARTLETT, M. S. and DIANANDA, P. H. [1] 'Extensions of Quenouille's test for autoregressive schemes', *J.R. Statist. Soc. B.*, **12** (1950) 108–115

BARTLETT, M. S. and RAJALAKSHMAN, D. V. [1] 'Goodness of fit tests for simultaneous autoregressive series', *J.R. Statist. Soc. B.*, **15** (1953) 107–124

BLACKMAN, R. B. and TUKEY, J. W. [1] 'The measurement of power spectra from the point of view of communications engineering Part 1', *Bell System Technical Journal*, **37** (1958) 185–282. 'Part 2', *Bell System Technical Journal*, **37** (1958) 485–569

COURANT, R. and HILBERT, D. [1] *Methods of Mathematical Physics.* Vol. 1, New York (1953)

CRAMER, H. [1] *Mathematical Methods of Statistics.* Princeton (1946)
[2] 'On the theory of stationary random processes', *Ann. of Math.*, **41** (1940) 215–230

DANIELS, H. E. [1] 'The approximate distribution of serial correlation coefficients', *Biometrika*, **43** (1956) 169–185

DIANANDA, P. H. [1] 'Some probability limit theorems with statistical applications', *Proc. Camb. Phil. Soc.*, **49** (1953) 239–246

DIXON, W. J. [1] 'Further contributions to the problem of serial correlation', *Ann. Math. Statist.*, **15** (1944) 119–144

DOOB, J. L. [1] *Stochastic Processes.* New York (1952)

DURBIN, J. and WATSON, G. S. [1] 'Testing for serial correlation in least squares regression I', *Biometrika*, **37** (1950) 409–428
[2] 'Testing for serial correlation in least squares regression II', *Biometrika*, **38** (1951) 159–177

FELLER, W. [1] 'On the Kolmogorov–Smirnov theorems for empirical distributions', *Ann. Math. Statist.*, **19** (1948) 177–189

FISHER, R. A. [1] 'Tests of significance in harmonic analysis', *Proc. Roy. Soc. A.*, **125** (1929) 54–59
[2] 'On the similarity of the distributions found for the test of significance in harmonic analysis and in Steven's problem in geometrical probability', *Annals of Eugenics*, **10** (1940) 14–17

GRENANDER, U. [1] 'On empirical spectral analysis of stochastic processes', *Ark. Mat.*, **1** (1951) 503–531
[2] 'On the estimation of the regression coefficients in the case of an autocorrelated disturbance', *Ann. Math. Statist.*, **25** (1954) 252–272

GRENANDER, U. and ROSENBLATT, M. [1] *Statistical Analysis of Stationary Time Series.* New York (1957)
[2] 'Comments on statistical spectral analysis', *Skand. Aktuarietidskr.*, **36** (1953) 182–202
[3] 'Statistical spectral analysis of time series arising from a stationary stochastic process', *Ann. Math. Statist.*, **24** (1953) 537–558

HALMOS, P. R. [1] *Measure Theory.* New York (1950)
[2] *Finite Dimensional Vector Spaces.* Princeton (1942)

HANNAN, E. J. [1] 'Exact tests for serial correlation', *Biometrika*, **42** (1955) 133–142

[2] 'Testing for serial correlation in least squares regression', *Biometrika*, **44** (1957) 57–66

[3] 'The estimation of the spectral density after trend removal', *J.R. Statist. Soc.*, **20** (1958)

[4] 'The asymptotic powers of certain tests of goodness of fit for time series', *J.R. Statist. Soc.* **20** (1958) 143–151

HART, B. I. [1] 'Tabulation of the probabilities for the ratio of mean square successive difference to the variance', *Ann. Math. Statist.*, **13** (1942) 207–214

HILLE, E. and PHILLIPS, R. S. [1] *Functional Analysis and Semi Groups*. American Mathematical Society Colloquium Publications Vol. XXXI (1957)

JENKINS, G. H. [1] 'An angular transformation for the serial correlation coefficient', *Biometrika*, **41** (1954) 261–265

[2] 'Tests of hypotheses in the linear autoregressive model I. Null hypothesis distribution in the Yule scheme', *Biometrika*, **41** (1954) 405–419

[3] 'Tests of hypotheses in the linear autoregressive model II. Null distributions for higher order schemes. Non-null distributions', *Biometrika*, **43** (1956) 186–199

JOWETT, G. H. 'The comparison of means of sets of observations from sections of independent stochastic series', *J.R. Statist. Soc. B.*, **17** (1955) 208–227

KENDALL, M. G. [1] *Contributions to the Study of Oscillatory Time-Series*. Cambridge (1946)

[2] *The Advanced Theory of Statistics*, Vol. 1, London (1945); Vol. II, London (1946)

[3] 'Tables of autoregressive series', *Biometrika*, **30** (1949) 267–289

KOLMOGOROFF [1] *Foundations of the Theory of Probability*. New York (1950)

[2] 'Sur l'interpretation et extrapolation des suites stationnaires', *C.R. Acad. Sci. Paris*, **208** (1939) 2043–2045

LEIPNIK, R. B. [1] 'Distribution of the serial correlation coefficient in

a circularly correlated universe', *Ann. Math. Statist.*, **13** (1947
80–87

LITTLEWOOD, J. E. [1] *Theory of Functions*. Oxford (1944)

LOMNICKI, Z. A. and ZAREMBA, S. K. [1] 'On the estimation of autocorrelation in time series', *Ann. Math. Statist.*, **28** (1957) 140–158

MORAN, P. A. P. [1] 'The statistical analysis of the Canadian lynx cycle I. Structure and Prediction', *Aust. J. Zoology*, **1** (1953) 163–173
[2] 'A test for the serial independence of residuals', *Biometrika*, **37** (1950) 178–181

MUNROE, M. E. [1] *Introduction to Measure and Integration*. Cambridge, Mass. (1953)

NEUMANN, J. VON [1] 'Distribution of the ratio of the mean square successive difference to the variance', *Ann. Math. Statist.*, **12** (1941) 367–395

OGAWARA, M. [1] 'A note on the test of serial correlation coefficients', *Ann. Math. Statist.*, **22** (1951) 115–118

OLKIN, I. and PRATT, J. W. [1] 'A multivariate Tcheybeycheff inequality', *Ann. Math. Statist.*, **29** (1958) 201–211

PARZEN, E. [1] 'On consistent estimates of the spectrum of a stationary time series', *Ann. Math. Statist.*, **28** (1957) 329–348

PITMAN, E. J. G. [1] 'The "closest" estimates of statistical parameters', *Proc. Camb. Phil. Soc.*, **33** (1937) 212–222

QUENOUILLE, M. H. [1] 'The joint distribution of serial correlation coefficients', *Ann. Math. Statist.*, **20** (1949) 561–571
[2] 'A large sample test for the goodness of fit of autoregressive schemes', *J.R. Statist. Soc. A.*, **110** (1947) 123–129
[3] *Associated Measurements*. London (1952)

RIESZ, F. and SZ-NAGY, B. [1] *Functional Analysis*. London (1956)

Royal Statistical Society Symposia on Time Series [1] 'Autocorrelation in time series', *J.R. Statist. Soc.* (*Suppl.*), **8** (1946) 27–97
[2] 'Spectral approach to time series', *J.R. Statist. Soc. B.*, **19** (1957) 1–63
[3] 'Analysis of geophysical time series', *J.R. Statist. Soc. A.*, **120** (1957) 387–439

SMIRNOV, N. [1] 'Tables for estimating the goodness of fit of empirical distributions', *Ann. Math. Statist.*, **19** (1948) 279–281

SOMMERFELD, A. J. W. [1] *Partial Differential Equations in Physics.* New York (1949)

TITCHMARSH, E. C. [1] *Introduction to the Theory of Fourier Integrals.* Oxford (1937)

[2] *The Theory of Functions.* Oxford (1932)

WALD, A. [1] 'Asymptotic properties of the maximum likelihood estimate of an unknown parameter of a discrete stochastic process', *Ann. Math. Statist.*, **19** (1948) 40–46

WALKER, A. M. [1] 'The asymptotic distribution of serial correlation coefficients for autoregressive processes with dependent residuals', *Proc. Camb. Phil. Soc.*, **50** (1954) 60–64

WATSON, G. S. [1] 'On the joint distribution of the circular serial correlation coefficients', *Biometrika*, **43** (1956) 161–168

[2] Serial Correlation in Regression Analysis. Thesis submitted to the Faculty of the North Carolina State College, 1951

[3] 'Serial Correlation in Regression Analysis I', *Biometrika*, **42** (1955) 327–341

WATSON, G. S. and HANNAN, E. J. [1] 'Serial Correlation in Regression Analysis II', *Biometrika*, **43** (1955) 436–448

WHITTAKER, E. T. and WATSON, G. N. [1] *A Course of Modern Analysis.* Cambridge (1946)

WHITTLE, P. [1] *Hypothesis Testing in Time Series Analysis.* Uppsala (1951)

[2] 'Estimation and information in stationary time series', *Ark. Mat.*, **2** (1953) 423–434

[3] 'The statistical analysis of a seiche record', *Sears Foundation Journal of Marine Research*, **13** (1954) 76–100

[4] 'Tests of fit in time-series', *Biometrika*, **39** (1952) 309–318

WOLD, H. [1] *A Study in the Analysis of Stationary Time Series.* Stockholm (1953)

[2] 'A large sample test for moving averages', *J.R. Statist. Soc. B.*, **11** (1949) 297–305

Index